Nanomaterials for Energy Applications 2014

Topical Conference at the 2014 AIChE Annual Meeting

Atlanta, Georgia, USA
16-21 November 2014

ISBN: 978-1-5108-1245-1

Printed from e-media with permission by:

Curran Associates, Inc.
57 Morehouse Lane
Red Hook, NY 12571

Some format issues inherent in the e-media version may also appear in this print version.

Printed by Curran Associates, Inc. (2015)

For permission requests, please contact AIChE
at the address below.

AIChE
120 Wall Street, FL 23
New York, NY 10005-4020

Phone: (800) 242-4363
Fax: (203) 775-5177

www.aiche.org

Additional copies of this publication are available from:

Curran Associates, Inc.
57 Morehouse Lane
Red Hook, NY 12571 USA
Phone: 845-758-0400
Fax: 845-758-2634
Email: curran@proceedings.com
Web: www.proceedings.com

TABLE OF CONTENTS

Silicon (Si) nanomaterials are leading candidates for lithium-ion battery anodes with an order of magnitude greater gravimetric and volumetric Li storage capacity relative to commercial graphite. Si nanowires are particularly compelling because they provide short Li diffusion paths due to their narrow diameter combined with long continuous paths for electron transport down their length. However, the low electrical conductivity of Si makes it necessary to add conductive additives, typically conductive carbon particles. This lowers the electrode charge storage capacity and can lead to poor performance due to the segregation of the carbon particles in the electrode layer during processing or as a result of mechanical stresses during cycling. Here, we show that stable battery performance can be achieved without the addition of conductive carbon particles by either encapsulating Si nanowires in a uniform carbon film during synthesis or by adding conductive tin to the Si nanowires.

Carbon encapsulation provides good adhesion to the Si material to improve cycle stability and electrical conductivity, and provides additional mechanical support against the stresses that occur during volume expansion and contraction. The carbon coating also creates a protective barrier that prevents unwanted side reactions between the Si surface and the electrolyte to form a robust solid electrolyte interphase layer. Tin-seeded Si nanowires with conformal carbon coatings were tested in lithium ion battery half-cells without any conductive additives, exhibiting capacities over 2000 mAh/g for 100 cycles when cycled at C/10 and over 1200 mAh/g when cycled more rapidly at 1C. In situ transmission electron microscopy (TEM) observation reveals that the carbon skin speeds lithiation of the Si nanowires significantly and constrains the final volume expansion.

Alternatively, we show that Si nanowires formed with significant tin incorporation can significantly increase electrical conductivity and lithiation/delithiation rates without the extra weight of carbon. Tin was found to incorporate into Si nanowires at concentrations up to 10 at% in tin-seeded supercritical fluid-liquid-solid (SFLS) Si nanowire growth reactions. Coin cell testing showed that without conductive additives the heavily tin incorporated Si nanowire anode films have a stable capacity of ~1000 mAh/g for over 100 cycles at a rate of 1C. This work shows that silicon-based lithium ion battery anode performance can be improved by tailoring the anode composition and structure of the nanomaterials.

390581 Direct Deposit of Polyvinyl Alcohol/Silicon/Graphene Nanoribbon Nanofibers for a Facile Production of High Capacity Lithium ion Battery Anodes

Abstract:

Despite high theoretical capacity in lithium-ion batteries, silicon (Si) anodes are still vulnerable to severe capacity fading due to dramatic volume expansion and formation of solid–electrolyte interface on their surfaces.[1,2] Herein, we prepared a novel anode of polyvinyl alcohol (PVA)/Si nanoparticles (NPs)/graphene nanoribbon (GNR) nanofibers directly on the current collector via water–based electrospinning. This facile direct deposit approach using water as solvent eliminates the use of toxic solvents like NMP and by–passes conventional paste preparation steps such as sonication, blending, grinding and coating. Furthermore, PVA polymer acts as a dispersant to disperse highly loaded GNRs in water, which are unzipped from carbon nanotubes. Battery cells using PVA/Si/GNRs nanofibers exhibit an initial discharge capacity over 5000 mAh/g at 0.1C due to the synergy between well–dispersed GNRs and Si NPs. Hybrid nanofibers also exhibit much less charge transport resistance and higher peak current density than Si NPs. Even at 1C, they show a capacity around 1800 mAh/g owing to highly conductive GNRs within nanofibers, and a very stable cycle retention over 90% during 200 cycles. Such excellent cycle retention is attributed to cross–linked PVA covering Si NPs, which not only prohibits the volume expansion but also alleviates the formation of SEI layers.

References

[1] H. Wu, Y. Cui, *Nano Today* **2012**, *7*, 414.

[2] Wu H.; Chan G.; Choi J. W.; Ryu I.; Yao Y.; McDowell M. T.; Lee S. W.; Jackson A.; Yang Y.; Hu L.; Cui Y.

Nature Nanotech. **2012**, *7*, 310.

384628 Foamed Mesoporous Carbon/Silicon Composite Nanofiber Anode for Lithium Ion Batteries

Abstract:

Lithium-ion battery is considered as one of the most promising energy storage and renewable power sources of the next generation. Si-based anode faces adverse issues on electrode pulverization and capacity loss because of the dramatic volume change and stress induction during frequent charging/discharging cycles. Such issues have been successfully mitigated when adopting silicon or carbon in the format of nanowires, or nanotubes, or mesoporous nanoparticles. However, the required synthesis processes are not suitable for scalable manufacturing. We here present a new mesoporous nanofiber Si/C composites synthesis process, in which mesoporous polymer nanofibers were first produced by electrospinning and foaming processes. Silicon nanoparticles are intercalated into carbon nanofibers during the spinning/foaming process. The mesopores and macrovoids inside carbon matrix provides the necessary free space at different scales to accommodate the volume expansion and the consequent stress within the electrode materials to minimize electrode pulverization during alloying process. The hybrid structure of macropores and mesopores also offers high surface-to-volume ratios and facilitate lithium transport between electrolytes and the active materials to achieve high charging/discharging rates. Our electrochemical tests demonstrated higher reversible capacity and better capacity retention with this porous carbon/silicon composite nanofiber anode when compared with that made of nonporous composite nanofibers and carbon nanofibers alone with similar treatment.

385016 A Mechanistic Understanding of Lithium-Ion Diffusion and Intercalation in Novel Lignin-Derived Carbon Composite Anodes

Abstract:

Novel lignin-based carbon composite materials consisting of amorphous and crystalline domains have been developed for use as anodes in lithium-ion batteries. The performance of these anodes is comparable to conventional anode materials at a significantly reduced manufacturing cost. However, the mechanism behind the performance of these novel materials is unknown. In this work, we develop an understanding of this mechanism through the use of reactive molecular dynamics simulations performed on computational models of the experimental systems. The coefficient of lithium-ion diffusivity in each domain and the degree of ion intercalation is ascertained as a function of ion-loading and related to structural properties of each composite system, including crystallite size, volume fraction of crystalline material, and composite density. Voltage profiles of the model systems are compared to those obtained from experiment. The understanding of lithium-ion diffusion within these systems is used to formulate a prediction of an ideal combination of material properties that would allow the development of even higher performing composite anodes.

N.M. was supported by a grant from the Oak Ridge Associated Universities High Performance Computing Program, by a grant from the Sustainable Energy Education and Research Center of the University of Tennessee and by a grant from the National Science Foundation (DGE-0801470). This research project used resources of the National Institute for Computational Sciences (NICS) supported by NSF under agreement number: OCI 07-11134.5. This research at Oak Ridge National Laboratory's Spallation Neutron Source was sponsored by the U.S. Department of Energy, Office of Basic Energy Sciences.

379944 Carbon-Based Nanostructured Electrodes for High Performance Rechargeable Battery Applications

Abstract:

Large-scale energy storage applications, such as electric vehicle, smart-grid, and renewable energy storage, drive the development of more sustainable and affordable carbon-based organic electrodes. Recently, it has been shown that oxygen functional groups on the surface of carbon materials, such as graphene oxide and carbon nanotube (CNT), can react with Li-ions reversibly at ~3 V vs Li, proposing these oxidized carbons as new rising positive electrodes for the Li-ion batteries. As the redox reactions primarily take place on the surface of the carbons, with no intercalation reactions involved, these electrodes generally show both high power and high energy density, and good cycling stability. In this work, we demonstrate high-performance nanostructured graphene oxide and CNT electrodes including the surface redox reaction between oxygen functional groups and lithium ions. The synthesized nanostructured carbon electrodes showed high gravimetric capacity of ~140 mAh/g in two electrode lithium-cells. We further demonstrate these nanostructured carbon electrodes are promising substrates for nano-scale coatings of various oxides, molecules, and polymers to design high-performance energy storage devices.

363568 A Facile Hydrothermal Route to Fe2O3 with Conductive Additives As Composite Anode for Lithium Ion Batteries

Abstract:

Transition metal oxides have been intensively investigated as anode materials for lithium ion batteries (LIBs) due to their high theoretical capacity, natural abundance, low cost, ease of fabrication and environmental friendliness. However, poor conductivity and large volume change of such materials during electrochemical reaction strongly limit their practical application. To overcome these disadvantages, research has been focusing on the following aspects: (i) synthesis of nanosized metal oxide materials with large surface area, which can greatly enhance the contact area between the active material and electrolyte and shorten the diffusion path of lithium ions; (ii) novel morphologies of the microstructures such as hollow sphere, nanorod, spindle, nanotube, nanocube proposed as means of advancing superior electrochemical kinetics and remarkable structural stability; and (iii) composites of metal oxides with carbon or polymer additives designed to improve the conductivity and the electrochemical performances.

Herein, we report a facile hydrothermal route for the synthesis of Fe_2O_3 nanoparticles and their composites with conductive additives, reduced graphite oxide (rGO) and conductive polymer of Poly(3,4-ethylenedioxythiophene) Polystyrene sulfonate (PEDOT:PSS), as anode materials for LIBs. The addition of conductive polymer layer on Fe_2O_3 nanoparticles may facilitate the electron transport but hinders the Li^+ ion insertion at higher current density. On the other hand, Fe_2O_3@rGO prepared by growing Fe_2O_3 nanoparticles directly on rGO sheets shows the best electrochemical behavior due to a beneficial combination of the rGO nanosheets partially wrapped Fe_2O_3 nanoparticles which facilitates both the electron transport and the Li^+ ion insertion. The enhanced conductivity of the composites was proved. The high specific capacity and stable rate performance of Fe_2O_3@rGO composites encourages their further application as potential candidate for the anode materials in LIBs. These results will be helpful in further elucidation of the role of conductive additives in improving the electrochemical performance of Fe_2O_3 based composite anodes and this simple synthetic strategy may be applied for the large scale production of metal oxides with conductive additives for LIBs.

380860 Advanced Hybrid Supercapacitor Based on New Ordered Mesoporous Materials As High Performance Anodes

Abstract:

Although the high energy-power performances with superior cyclability have been achieved in hybrid supercapacitors (HSCs), the development of advanced HSCs have recently stagnated. This is mainly because the anode materials of HSCs were very limited to the titanium-based oxides such as $Li_4Ti_5O_{12}$ and TiO_2. Herein, we developed advanced hybrid supercapacitor based on new mesoporous metal oxide andoes. The resulting hybrid capacitor delivered high energy and power densities with exceptional cycle stability, which is comparable to, or even higher than, those of state of the art Li-ion battery and supercapacitor systems.

379782 Ultra-Thin Polythiophene within Nanostructured Electrode with Enhanced Charge Storage Capacity

Abstract:

Unsubstituated polythiophene (PTh) was synthesized using a unique oxidative chemical vapor deposition (oCVD) approach. The deposited polymer was found to be conductive with measured conductivity as high as 70 S/cm, which is an indication of a low degree of defects. The deposited PTh film showed strong electrochromic activity as evidenced by the color change from dark blue to bright red upon charge-discharge in organic electrolyte solutions. The charge storage capacity was measured in a three electrode setup using 1 M tetraethylammonium tetrafluoroborate in acetonitrile and values up to 70 F/g were obtained. Polymer films of thickness above 800 nm were found to be unstable and degraded upon charge-discharge as evidenced by the formation of micro-cracks and loss of capacity. However, thinner films with thickness down to 500 nm were stable upon charge-discharge. The observed charge trapping in the film at high scan rates (>200 mV/s) disappeared when the film thickness was reduced further to sub-200 nm. The electrochemically active film was also successfully deposited within mesoporous electrodes. To investigate the influence of nanostructure on electrochemical charge storage capacity of the polymer, the parameters influencing oCVD were carefully adjusted to achieve conformal coating within the nanostructure of anodized aluminum oxide with straight pores of ~200 nm in diameter and mesoporous TiO_2 consisting of random 10-25 nm pore channels within a sintered network of TiO_2 nanoparticles The polymer deposited within the 3D TiO_2 network and assembled on a conductive electrode was characterized in a three electrode setup for charge storage capacity. The charge storage capacity of the bare, uncoated titania electrode matrix was measured to be around 1 F/g, while the coated sample showed a capacitance as high as 250 F/g. Additionally, the specific capacitance of the polymer coating within the nanostructured electrodes was compared with that deposited on planar electrodes as blanket films. Interestingly, we have realized a further increase in the specific capacitance when the polymer was integrated as an ultrathin layer within the TiO_2 electrode compared to the blanket film. The enhanced capacitance was found to be not related to any mass transfer limitations within the different polymer films. We attribute the enhanced charge storage capacity to a nanoconfinement effect of the polymer within the nanopores. The enhanced capacitance, which is concomitant with the increase in the peak height in the cyclic voltammogram, suggests that the charge stored per unit of monomer exceeds that observed on the planar electrode. This observation led us to fabricate electrodes by using conventional activated carbon as the matrix layer. For coatings on activated carbon, a polymer-to-activated carbon mass ratio of ~1.5 was found to deliver a maximum specific capacitance that is ~50% higher than that of bare activated carbon, 145 vs. 92 F/g. This capacitance translates to over a 250% increase in volumetric capacitance since the volume contribution of the ultrathin polymer coating is negligible (120 vs. 47 F/cm^3).

366892 Roll-to-Roll Synthesis of Vertically Aligned Carbon Nanotube Electrodes for Electrical Double Layer Capacitors

Abstract:

Electrochemical double layer capacitors (EDLCs), also known as supercapacitors, have received notable attention in recent years for applications requiring durable and reliable devices that can deliver high power output with high energy density. EDLCs provide energy densities 2 to 3 orders of magnitude greater than that of electrolytic capacitors, while offering comparable power densities, thus, providing a suitable candidate power systems requiring a combination of power and energy. EDLCs have been commercialized using activated carbon electrodes, primarily due to its high surface area, high porosity and electrochemical stability.

Vertically aligned carbon nanotubes (VACNTs), however, have emerged as an alternative material for EDLCs due to their high conductivity, electrochemical stability, facile synthesis and the ease of controlling the ion-accessible area by varying CNT density on substrates. Although CNTs can be synthesized in large quantities, present processes are not amenable to continuous synthesis of VACNTs directly on current collectors for scalable manufacturing of EDLC electrodes, which is necessary to maintain low costs for commercialization. Several methods have been investigated for CNT synthesis, including electric arc discharge, laser ablation and chemical vapor deposition (CVD). Of these, CVD has emerged as a practical and reliable method for synthesizing VACNT forests with tunable properties (e.g., tube diameter, number of walls, and dopant ratio). During material synthesis, three primary factors limit continuous manufacturing of nanomaterials: i) substrate size determined by reactor geometry, ii) requirements of complex catalytic substrate preparation, and iii) high operating temperatures that are incompatible with traditional current collectors (e.g., Al foil).

In this work, we present a scalable, roll-to-roll process for synthesizing forests of VACNTs on inexpensive Al foil substrates using the ferrocene-xylene liquid injection floating catalyst technique for the continuous production of CNT based EDLC electrodes. The Al foil is continuously drawn through a chemical vapor deposition reactor operating at ambient pressure and a relatively low growth temperature (600 °C). EDLC electrodes comprising VACNT forests synthesized in continuous and stationary CVD processes were directly assembled into supercapacitor cells, which yielded high power densities (1270 W/kg) and energy densities (11.5 Wh/kg). These devices exhibited very low internal series resistance due to their intimate contact with the current collector and excellent cycle stability with no loss in performance over multiple thousands of cycles.

375467 In Situ One-Step Synthesis of Hierarchical Nitrogen-Doped Porous Carbons Derived from Metal-Organic Frameworks for Supercapacitors

Abstract:

Undoped porous carbon is a widely explored electrode material for supercapacitors due to its low cost, good stability, and high surface area. However, undoped porous carbon shows limited performance since it stores charge by a single mechanism *via* the electric double layer. Recently, nitrogen-doped porous carbon was found to exhibit enhanced electrochemical performance due to its altered chemical structure and additional pseudocapacitance originating from nitrogen heteroatoms. To synthesize nitrogen-doped porous carbons various preparation methods have been explored such as chemical vapor deposition, plasma treatment, thermal annealing, etc. Most of these methods require multiple steps and are not easily scaled. Furthermore, it is very challenging to control nitrogen content.

Here, we report the synthesis of nitrogen-doped porous carbon by one-step carbonization of isorecticular metal-organic frameworks (IRMOF-3). No additional template or pore-forming agent, or carbon and nitrogen sources are required. IRMOF-3 plays the role of both carbon and nitrogen sources as well as a self-template. Furthermore, the nitrogen content can be fine-tuned by changing carbonization temperature. The synthesized nitrogen-doped porous carbon exhibits excellent capacitance as high as 239 F g^{-1}, which is much higher than analogous nitrogen-free porous carbons.

Highly Electrocapacitive Nitrogen-Doped Graphitic Porous Carbons for Supercapacitor Electrodes

Kyung Taek Cho[1], *Sang Bok Lee*[2], *and Jae W Lee*[1]*

1Department of Chemical and Biomolecular Engineering, KAIST, Daejeon, 305-701, Korea.

2Department of Chemistry and Biochemistry, University of Maryland College Park, MD 20742, USA.

Nitrogen-doped graphitic porous carbons (NGPCs) were produced through polyacrylonitrile (PAN) with two-step heating processes. The PAN has used for carbon fibers in the industrial fields because of its low-cost, outstanding physical and electrical properties. Although porous carbons made from PAN have frequently been researched, the methods required some special equipment and tedious processes. Here, The NGPCs are obtained by the stabilization of PAN at 290°C and the subsequent carbonization with the KOH activation at 700 – 800°C. The porous carbons had high specific surface area (above $3000 m^2 g^{-1}$), evenly distributed pores (0.5-5nm) and graphitic structures. Based on these characteristics, in addition to nitrogen doped in carbons, NGPSs can be substituted for carbon materials obtained from conducting polymer in various applications. This work will illustrate that NGPCs retain high, reproducible capacitance in the application supercapacitors electrodes.

For energy storage, supercapacitors can provide high power density and long cyclability. However, the energy density needs to improve to make them economical for portable electronics and hybrid electric vehicles. The addition of pseudocapacitance material to act as a Faradaic charge storage mechanism can increase the charge storage capability by up to two orders of magnitude. Transition metal oxides with multiple valence states are often used to add pseudocapacitance to supercapacitor electrodes. For this work, we used atomic layer deposition (ALD) to apply coatings of pseudocapacitance vanadium oxide to porous carbon powders. ALD produced conformal coatings that were controlled with nano-scale precision. Multiple types of carbon powders with varying surface area, pore structure, and conductivity were treated with various acid regimens to modify its chemical and physical properties. Reactive groups formed on the carbon, including esters, ethers, alcohols, ketones were characterized using FTIR spectroscopy before coating. The carbon powder's graphitic nature, surface area, porosity, and elemental content was characterized using Raman spectroscopy, Brunauer–Emmett–Teller (BET) analysis, scanning electron microscopy (SEM), and energy dispersive spectroscopy (EDS) respectively before and after applying the ALD vanadium oxide. X-ray photoelectron spectroscopy (XPS) and X-ray diffraction (XRD) showed that amorphous V_2O_5 was deposited on the carbon powder. The V_2O_5 ALD reduced the surface area by an average of 50%, thereby decreasing the available area for charge to be electrostatically stored in a Helmholtz double layer. The powder that underwent rigorous acid treatment showed twice the oxygen content, which aided in the nucleation of vanadium oxide on the surface of the powder, and lead to a 15 fold increase in vanadium loading when coated with ALD. The coated powder was then incorporated into electrodes and tested with cyclic voltammetry and electrochemical impedance spectroscopy to develop an understanding of the vanadium oxide's effect on supercapacitor electrode performance. The fundamental structural characterization of ALD coated carbon powders will facilitate the design and construction of supercapacitor electrodes.

387600 Surface and Structure Engineered Graphenes for Liquid-Phase Processing Toward High-Performance Porous Solid Electrode Films

Abstract:

Graphene has become a class of materials that have potential impact on diverse applications such as energy storage (supercapacitors or batteries), electronics, and membranes. Our studies consider the synthesis and processing route to produce structure controlled graphene particulate film of high-performance electrode, starting with low-cost graphite powders. This presentation will present a new chemical processing route with serial of step intermediate products: from graphite powders, graphene oxide (GO) sheet particles, surface modified GO particles, controlled self-assembly particles of GO or reduced GO (rGO) that contains the desirable sheet-stacking architecture, liquid dispersion of structure-engineered rGO particles, and to solid film deposit containing desirable pore size and architecture. Surface modification with the appropriate molecular ligands are used as key strategies to re-assembly the engineered graphene particles and controlled dispersion of graphene particles into either aqueous or non-aqueous solvents. We are reporting our recent results on overcoming the dispersion issue of graphene into non-polar solvents. We also show the significant capacitance improvement due to molecular ligand-assisted interspace control between graphene sheets inside each re-assembled or stacked particulate. Finally, we demonstrate that the pore size and pore structure control of particulate electrode film deposits is critical to the final film performance.

Double Layer and Pseudocapacitive Charge-Storage Mechanisms in Carbides and Nitrides

Abdoulaye Djire[a], *Olabode T Ajenifujah*[a], *Alice E. S. Sleightholme*[a], *Paul Rasmussen*[ab], *Levi T Thompson**[ab]

[a] Department of Chemical Engineering
[b] Hydrogen Energy Technology Laboratory
University of Michigan, Ann Arbor, MI 48109-2136, USA

Tel. (734) 936-2015
E-mail: ltt@umich.edu

Introduction

Early transition-metal carbides and nitrides are promising candidates for use in supercapacitor electrodes due to their high electronic conductivities, high surface areas (can exceed 200 m^2g^{-1}), good electrochemical stabilities and high capacitance [1, 2]. For example, the capacitance for VN has been reported to be as high as 1340 Fg^{-1} in aqueous KOH [3]. This high capacitance has attributed to a combination of electric double-layer formation and faradaic reactions occurring on the nitride or oxynitride (VN_xO_y) surface [3]. Despite efforts to date, the nature of the faradaic redox reactions or pseudocapacitive properties of early transition-metal carbides and nitrides remains ill-defined. This presents a challenge to the full exploitation of these materials. Here we report a detailed investigation of the charge-storage mechanisms in early transition-metal carbides and nitrides in aqueous media. The contributions of both double-layer and pseudocapacitive mechanisms have been deconvoluted using a combination of electrochemical techniques including cyclic voltammetry (CV), electrochemical impedance spectroscopy (EIS) and electrochemical quartz crystal microbalance (EQCM), x-ray absorption spectroscopy and neutron scattering.

Experimental

High-surface-area Ti, V, Nb, Mo, and W carbides and nitrides were prepared from their oxide precursors TiO_2 (Alfa Aesar), V_2O_5 (Alfa Aesar), N_2O_5 (Alfa Aesar), $(NH_4)_6Mo_7O_{24}.4H_2O$ (81-83% as MoO_3, Alfa Aesar) and WO_3 (Alfa Aesar) respectively, by temperature-programmed reaction (TPR) synthesis with 15% CH_4 / H_2 (Cryogenic Gases) or NH_3 (Cryogenic Gases), respectively, then passivated using a flowing mixture of 1% O_2/He (Cryogenic Gases). Characterization of the structural properties was performed using nitrogen physisorption (BET surface area) and X-ray diffraction. The CV was used to establish the stability windows and capacitances for these materials. The capacitance was deconvoluted into double-layer capacitance and pseudocapacitance by means of CV and EIS. EQCM was used to characterize the nature of the adsorbed/desorbed species during charge/discharge. For selected materials details regarding key species and hydrogen adsorption were determined from the x-ray absorption and neutron scattering results.

Results and Discussion

Figure 1 shows the response of VN in the frequency range of 10 kHz to 10 mHz in acidic medium for selected potentials within the stable potential window. We observed plateaus at -0.7 and -0.44V bias potentials. This region is believed to be a signature of double-layer and surface adsorption, respectively. This behavior is expected given that both processes depend strongly on accessible surface area. As shown in Figure 2, the double-layer capacitance determined for VN in acidic medium was approximately 96 μFcm^{-2}. The charge-storage mechanism was found to be a combination of surface redox reaction, adsorption and double-layer charging. Similar examinations have been applied to the other carbides and nitrides listed above in aqueous media and the results will be discussed.

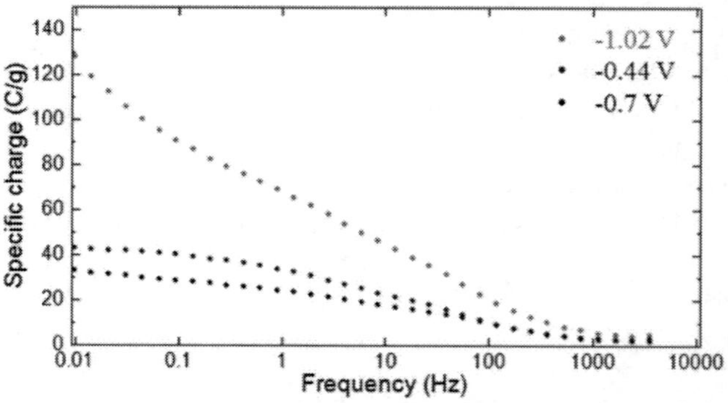

Figure 1: Plot of specific charge vs. frequency of VN in 0.1M H_2SO_4 at selected potentials.

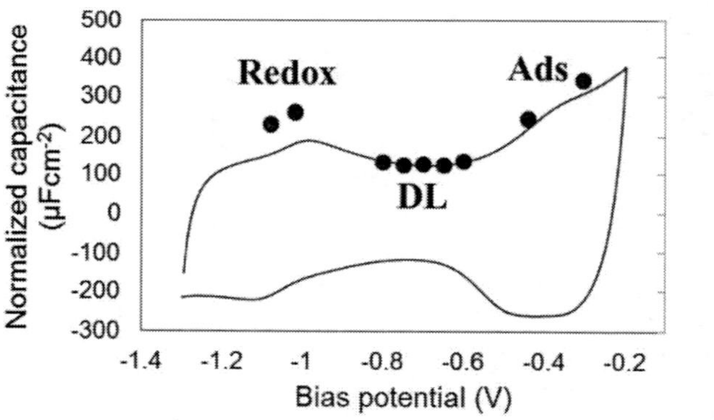

Figure 2: Plot of CV and capacitances (calculated from EIS) of VN in 0.1M H_2SO_4 showing the regions of surface redox reaction (Redox), double-layer (DL) and adsorption (Ads) within the stable window.

References

(1) Cladridge, J. B.; York, A. P. E.; Brungs, A. J.; Green Malcolm L. H.; *Chem. Mater.* **2000**, 12, 132.
(2) Wixom, M.R.; Tarnowski, D. J.; Parker, J. M.; Lee , J.Q.; Chen, P. –L.; Song, I.; Thompson, L. T.; *Mat. Res. Soc. Symp. Proc.* **1998,** 496, 643.
(3) Choi, D.; Kumta, P. N.; *Electrochem. Solid-State Lett.* **2005,** 8, 8, A418.

390123 Spherical Carbon Nanotube Assemblies and Their Supercapacitor Application

Abstract:

Macroscopic morphology of carbon nanotubes (CNTs) was designed using emulsion droplet confinement. Spherical CNT assemblies, were produced by the evaporation of CNT-dispersed aqueous emulsion droplets in oil. CNT was packed by the capillary pressure. With changing the concentration of CNT dispersion, the size of spherical CNT assemblies could be controlled. Average size of spherical CNT assemblies ca. 10 μm with BET specific area of $200 m^2/g$. The spherical CNT assemblies were heat-treated to remove residual solvent. When heat-treat at high temperature, the surface oxygenated groups were decomposed. CNT spherical particle-assembled film was applied to a supercapacitor. The specific capacitance of 500°C treated sample was 80F/g, a higher temperature decreased the capacitance with decreasing the pseudo-capacitance. Comparing with CNT films, the spherical CNT particle electrode showed 40% higher specific capacitance with higher rate performance, attributed to the compact packing of CNTs and spherical particles and the hierarchical porous structures in the assemblies.

385698 Ionic Liquids: Platforms for Thermally-Responsive Polymer Electrolytes

Abstract:

In recent years, remarkable progress in the development of batteries, supercapacitors and Li-ion technology has led to devices that deliver new records in power and energy density. The thermal instability in batteries, particularly lithium ion, creates a major roadblock to implementing these devices into large-format systems to support emerging, intermittent energy sources such as wind and solar. Efforts to mitigate safety issues such as flammability, reactivity, and thermal runaway typically involve the implementation of low conductivity solid-state materials or irreversible safety devices, which prove problematic for efficient, large-scale energy systems. Before high performance devices can be applied to large format energy storage applications, thermal safety must be improved.

"Smart materials" or stimuli-responsive materials are those with properties responding in a desired manner to an external stimulus or environmental change. While several classes of responsive materials exist, soft materials, particularly synthetic polymers, represent an opportunity to design systems with unique functionality to achieve a wide variety of chemical and physical properties. Here we present a system utilizing thermally-responsive polymers capable of reducing lithium conductivity or lithium charge transfer ability in non-aqueous electrolytes with increasing temperature. These proposed systems would create self-limiting lithium reactions within a battery should the temperature increase beyond a target, thus inhibiting thermal runaway.

Li-ion systems utilizing polymer gel electrolytes are commercially available and ionic liquids (ILs) have gained attention as potential replacements for traditional electrolytes (low volatility, non-flammability, high conductivity, and a wide potential window). Poly(ethylene oxide) (PEO) gels have long been combined with lithium salts to create polymeric electrolytes, which while safe, possess diminished conductivity. These PEO systems have recently been combined with ILs to boost performance while retaining the safety and mechanical stability of conventional polymer gel electrolytes. In our proposed system, we combine current research efforts utilizing polymer gel electrolytes and ILs, both of which provide improved thermal stability, while making use of the recently discovered lower critical solution temperature (LCST) between PEO and certain ILs, to modulate electrolyte conductivity with temperature [1,2]. We design the polymer electrolyte to decrease conductivity during a phase transition at elevated temperatures. As the temperature cools, the initial conductivity is reversibly restored as the electrolyte returns to its initial state. Additionally, we show how the PEO/IL composition and the addition of lithium salts affect both the magnitude of change in conductivity and the LCST in these thermally-responsive polymer electrolyte systems.

We then demonstrate how these thermally-responsive polymer electrolytes can be used to control electrochemical properties in both model supercapacitor systems, consisting of activated and mesoporous carbon supercapacitors, and lithium ion battery systems, consisting of graphitic anodes and lithium cobalt oxide ($LiCoO_2$) cathodes. Through the thermal phase transition above the normal operation temperatures of these electrochemical devices, we demonstrate how PEO/IL systems can be utilized to reduce the conductivity in electrolyte systems and reduce both the double layer charging ability in supercapacitor and redox charge transfer ability in lithium ion batteries (through a decrease in power and energy density).

The experiments performed to date show tremendous promise for controlling battery and supercapacitor operation using temperature. Further development of responsive electrolytes will lead to numerous opportunities associated with thermal safety and "smart" electrochemical systems that will ultimately lead to large-format battery applications will built-in thermal control that is both reversible and non-destructive.

1. H.-N. Lee and T. P. Lodge, J. Phys. Chem. Lett., 1, 1962 (2010).
2. H.-N. Lee, N. Newell, Z. Bai and T. P. Lodge, Macromolecules, 45, 3627 (2012).
3. T. Ueki and M. Watanabe, Langmuir, 23, 988 (2006).
4. K. Kodama et al., Langmuir, 25, 3820 (2009).
5. H.-N. Lee and T. P. Lodge, J. Phys. Chem. B, 115, 1971 (2011).

394219 Reverse Electrodialysis: Sustainable Energy from Hydraulic Fracturing Water Recycle

Abstract:

Due to the growing limitations in dependence on fossil fuels, attention has been turned towards searching for alternative means for harvesting energy. One emerging source is energy produced from natural salinity-gradients between freshwater and seawater. Two different processes used to accomplish this are pressure-retarded osmosis and reverse electrodialysis. Significant amounts of research are currently under way regarding pressure-retarded osmosis, however very few studies have been conducted on reverse electrodialysis due to heavy resistance in the diluate chamber of a reverse electrodialysis cell. The focus of this research is to reduce this resistance by using wafers and ion-exchange beads in order to maximize net power gain. Surrogate brackish/saltwater feed streams were used to measure overall power density.

By reducing resistance to levels previously deemed impossible, we are exploring the fundamental limits of separation technology. This research proposes an innovative resolution to high resistance levels through the use of ion exchange wafers to promote ionic movement. Here we show that such wafers have a great impact on the reverse electrodialysis process, and succeeded in achieving a sizeable gain in power density.

Traditional applications include harnessing energy from areas where freshwater bodies meet saltwater, as in an estuary. The future of reverse electrodialysis, however, lies in fracking. Produced water from a fracking well site in Oklahoma was used for power density calculations under realistic conditions. Preliminary results show that our system is capable of producing power densities of up to $10W/m^2$, which are five times higher than the current state-of-the-art. Environmental concerns will be all but eliminated if fracking can evolve to be a sort of cyclic process, where brackish fracking water is converted to energy to, in turn, power the overall operation. This poster presentation will model the power increase of electrode ionization and comment on future potential for improvement, especially in the realm of hydraulic fracturing applications.

394215 Modeling of the Aerobic Cometabolic Transformation of Chlorinated Ethenes By the Mycobacterium Elw-1

Abstract:

ELW1 is a recently isolated aerobic bacterium, in the mycobacterium family, that uses isobutene as a growth substrate. ELW1 was also discovered to have the ability to cometabolically transform substances with chemical structures similar to isobutene, such as chlorinated ethenes, into chloride ion and water. Chlorinated ethenes are notorious carcinogens, and groundwater pollutants, whose remediation has been a worldwide problem for over 40 years. Bioremediation of chlorinated ethenes by the augmentation of contaminated soils with microorganisms such as ELW1 is a proposed method of reducing contaminant concentrations in groundwater. To determine the viability of using ELW1 for groundwater remediation, an initial kinetic analysis of transformation rates must be performed. The focus of this research was to determine the rates of contaminant transformation when this organism was exposed to high concentrations of chlorinated ethenes such as trichloroethylene (TCE), *cis*-dichloroethene (c-DCE), and vinyl chloride (VC).

The ELW1 culture was grown and sustained in both batch culture and steady-state chemostat environments with isobutene as a growth substrate on a minimal growth media. Cells that were used for experiments were concentrated by centrifugation, washed with excess media and placed in a new media solution containing high concentrations of the compound to be tested. Initial rates of transformation were determined using gas chromatography to analyze 100μL samples of headspace gas over a period of one to three days to determine the rate of compound loss. The rates at which ELW1 can utilize isobutene were compared with rates of chlorinated ethene transformation, which overall quantifies the practicality of using this organism bioremediation purposes in the field.

Kinetic parameters of these cometabolic processes were determined using an analytical equation based on a modified Monod equation in order to account for decrease in cell activity over time. Results of this investigation indicate that the ELW-1 culture would be effective for the bioremediation of c-DCE and VC, but would be less effective for treated TCE.

394214 The Enzymatic Hydrolysis of Alfalfa Stalks for Use As a Biofuel Resource

Abstract:

Most biofuel resources, especially plant sources, have two key issues: farming space and slow growth. After overexpressing the sucrose phosphate synthase gene (SPS) in Medicago sativa (alfalfa) plants, large differences in phenotype and growth rate were seen in the plant. While the leaves of the alfalfa plants can be used for animal forage, the stems of the alfalfa can be fractionated into sugars that can then be fermented into bioethanols. This technique could solve the growth problem faced by bioethanol producers.

The purpose of the project was to fractionate and analyze the extractability of fermentable sugars from a control and several transgenic species that show overexpression in the SPS gene. Using a strong acid pre-treatment and enzymatic hydrolysis to break down the cellulose into glucose, the alfalfa stalks were fractionated then analyzed using ultraviolet spectroscopy with anthrone as an indicator for glucose content. Two of the three tested transgenic species were shown to have more glucose content per gram than the control species (2-4% more glucose/gram). A further development of the overexpression technique can lead to faster growth, increased concentration of fermentable carbohydrates, and larger crop yields.

UNIVERSITY OF SOUTH CAROLINA
College of Engineering and Computing

SAGE
Power through Engineering

Synthesis and Characterization of FeAlCr via High Throughput Techniques

Randy Fang
Jonathan Kenneth Bunn and Dr. Jason Hattrick-Simpers
Department of Chemical Engineering
University of South Carolina
AICHE National Conference 2014
11/17/2014

UNIVERSITY OF SOUTH CAROLINA
Honors College

1

Fukushima Daiichi Nuclear Disaster

http://digginintheclay.files.wordpress.com/2011/03/5525887859_e1934af238_o_d1.jpg

https://coto2.files.wordpress.com/2011/03/2-fuel-pellet-assembly.jpg

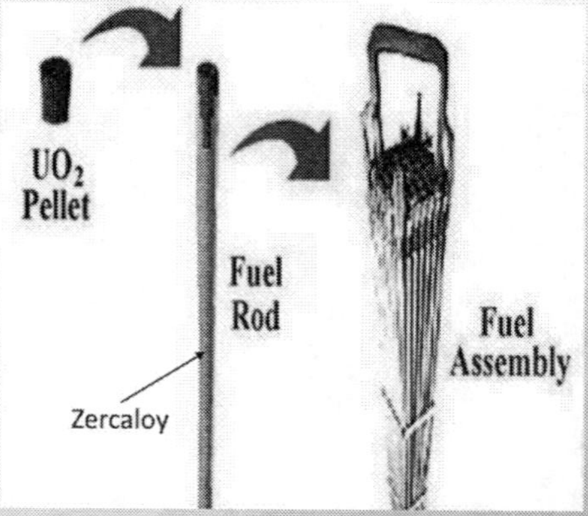

2

FeAlCr Alloy

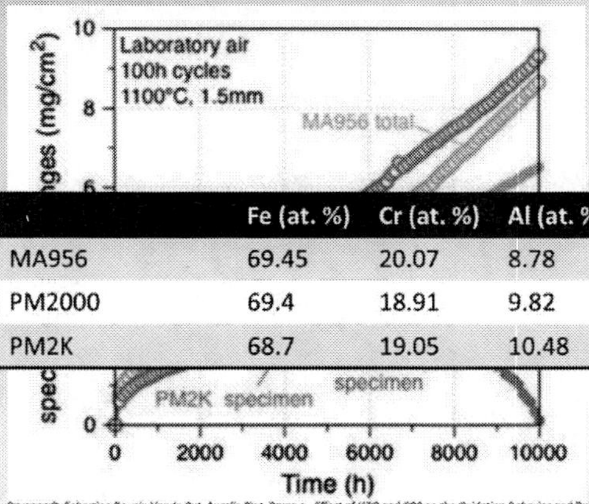

	Fe (at. %)	Cr (at. %)	Al (at. %)
MA956	69.45	20.07	8.78
PM2000	69.4	18.91	9.82
PM2K	68.7	19.05	10.48

Dryepondt, Sebastien Rouaix-Vande Put, Aurelie Pint, Bruce a., *Effect of H2O and CO2 on the Oxidation Behavior and Durability at High Temperature of ODS-FeCrAl*, Oxidation of Metals, Vol 79,2003

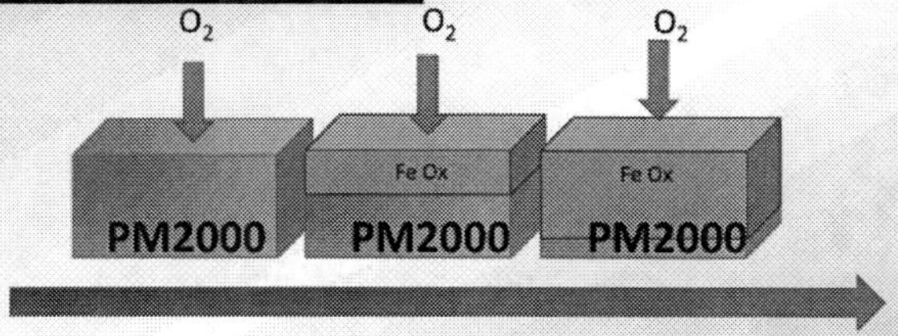

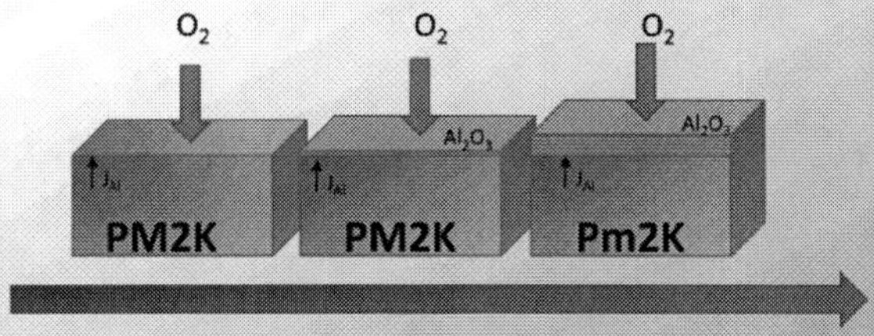

Introduction to High Throughput Techniques

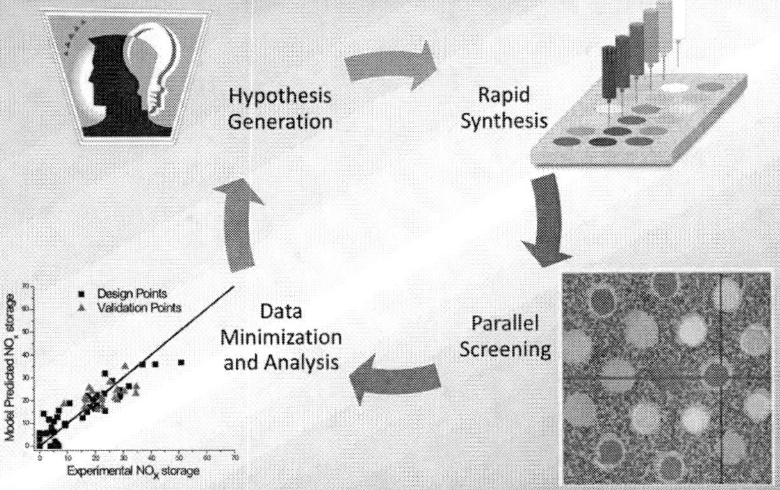

Research Objective

- **Implementing established high-throughput methods for FeAlCr high temperature oxidation resistance screening**
- Non-destructive characterization
- Gain comprehensive understanding of:
  - Time dynamics of oxide formation at high temperature
  - Role of metallurgical phase in promoting growth and continued growth of passivating oxide
- Identification of promising compositions for passivating oxide growth

5

High Throughput Approach to Explore the Ternary Phase Diagram

Qu Materials Science and Engineering A Vol 562 (2013)

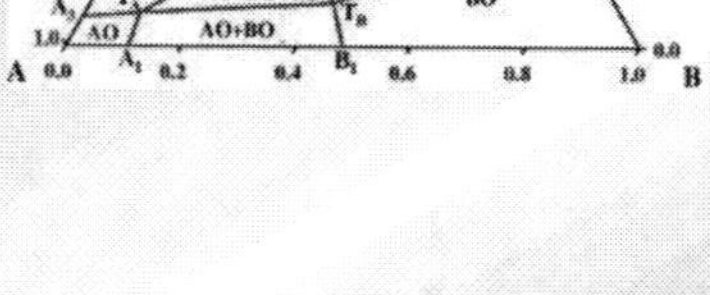

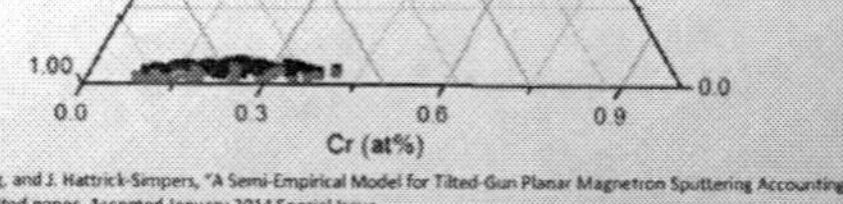

J. K. Bunn, C. J. Metting, and J. Hattrick-Simpers, "A Semi-Empirical Model for Tilted-Gun Planar Magnetron Sputtering Accounting for Chimney Shadowing ", JOM, invited paper, Accepted January 2014 Special Issue

6

Sputtering Geometry and Sample Locations

Ternary Deposition

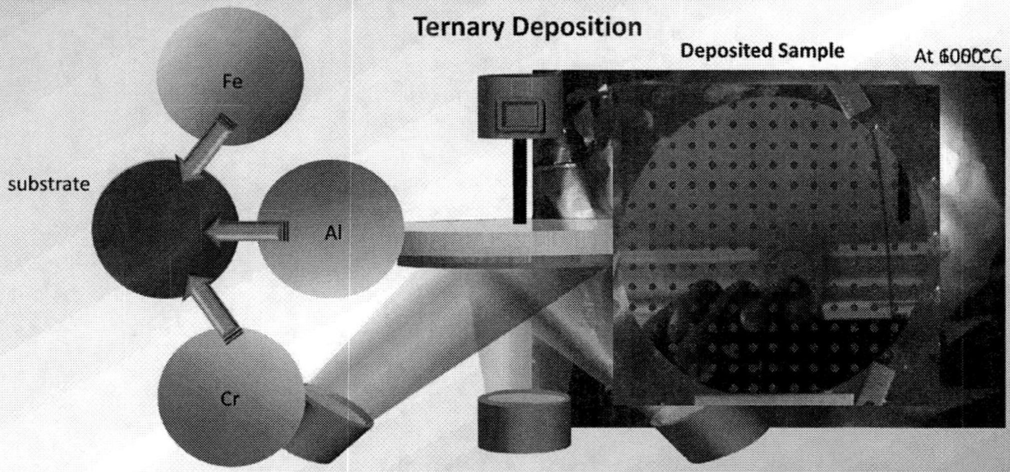

Magnetron Sputtering

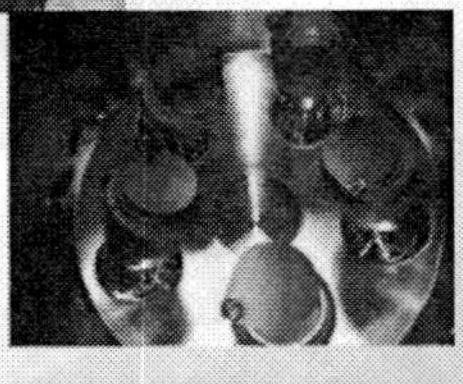

- **5 Guns (only used 3)**
- **DC/RF power sources**
- **Ultra high vacuum**
- **Deposition temperature at room temperature**

Experimental Design for High Temperature

Sample fabrication
-12 FeAlCr samples prepared on Al_2O_3 substrate and Ti with adhesion layer
-1 micron nominal thickness

After deposition

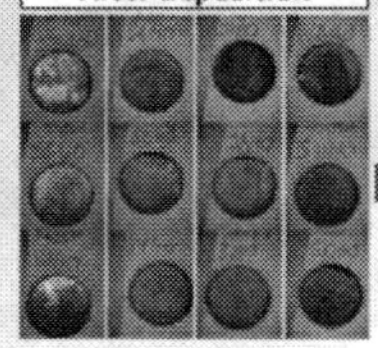

Characterization
-SEM/EDS
-XRD

Heat treatment
- Tube furnace treated
- Treated at 1050˚C
- Air quenched

Cycled 7 times for cumulative heat treatments at 1050°C for:

-5 min	-10 min
-20 min	-40 min
-60 min	-120 min
-240 min	-360 min

After heat treatments

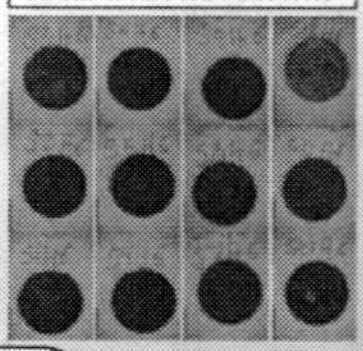

Characterization Techniques
- **Raman**: Fe_2O_3, Fe_3O_4, and Al_2O_3
- **XRD**: FeCrAl, Surface Oxides, and the substrate
- **Optical Microscopy**

9

Oxide Formation at 5 minutes, 1050˚C

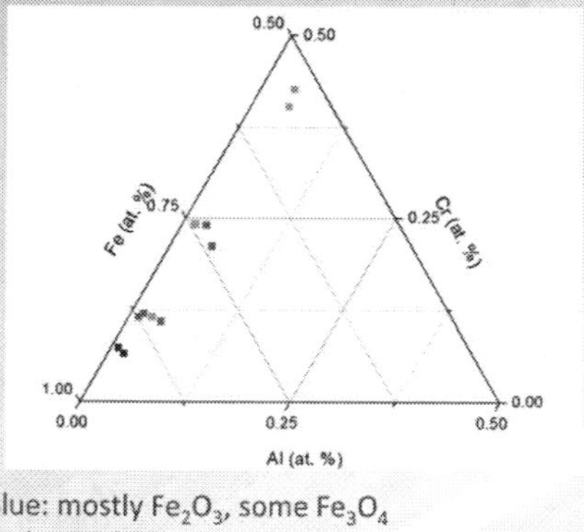

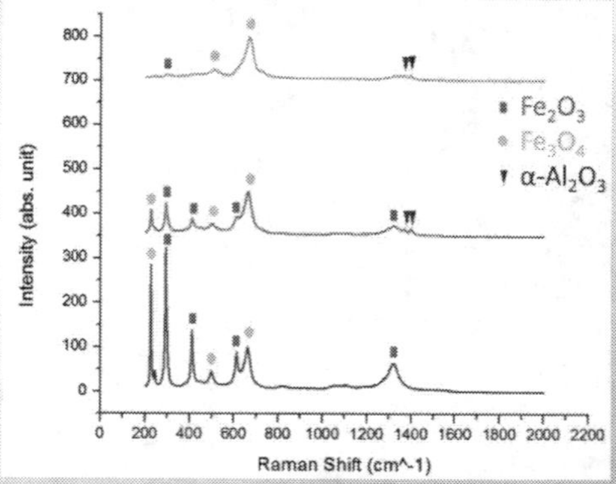

Blue: mostly Fe_2O_3, some Fe_3O_4
Red: mostly Fe_3O_4, some Fe_2O_3, some α-Al_2O_3
Green: very little Fe_2O_3, some α-Al_2O_3, some Fe_3O_4

- No α-Al_2O_3 growth was observed when the composition is Fe heavy

10

Oxide Formation at 360 minutes, 1050˚C

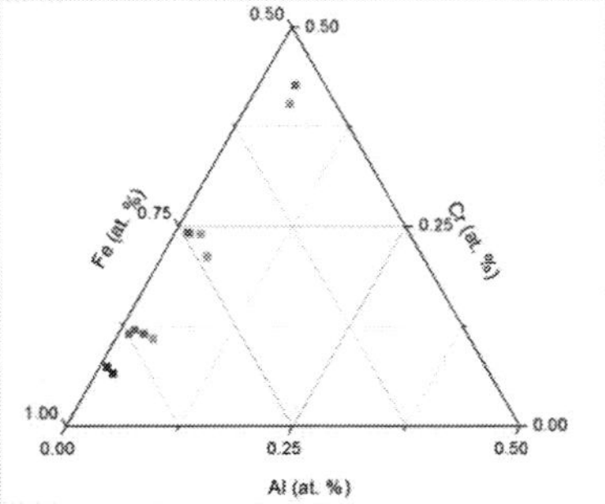

Blue: some Fe_2O_3 and Fe_3O_4, very little α-Al_2O_3
Red: some α-Al_2O_3, some Fe_3O_4
Green: majority α-Al_2O_3, some Fe_3O_4

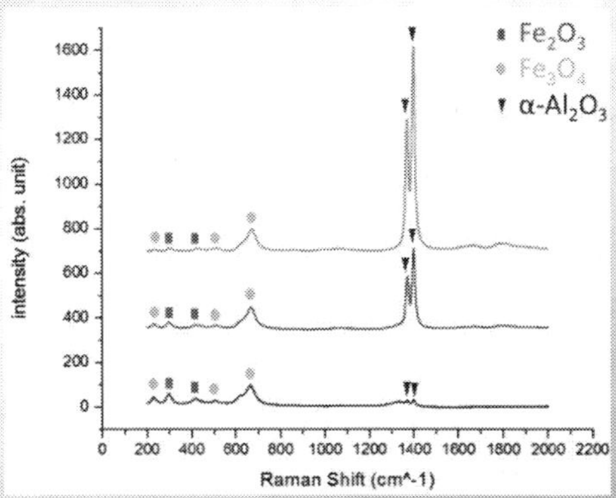

- Large amount of α-Al_2O_3 growth was observed at the Cr heavy region, indicating that Cr promotes the growth of the alumina oxides

Experimental Design at Low Temperature

Sample fabrication
-FeAlCr samples prepared on Si substrate with Ti diffusion barrier
-1 micron nominal thickness

Cycled 6 times for cumulative heat treatments 1 hour each (with the exception of 600°C):

-room temperature
-200°C
-300°C
-400°C
-500°C
-600°C (for 2hr)

Characterization
-*in-situ* GIXRD

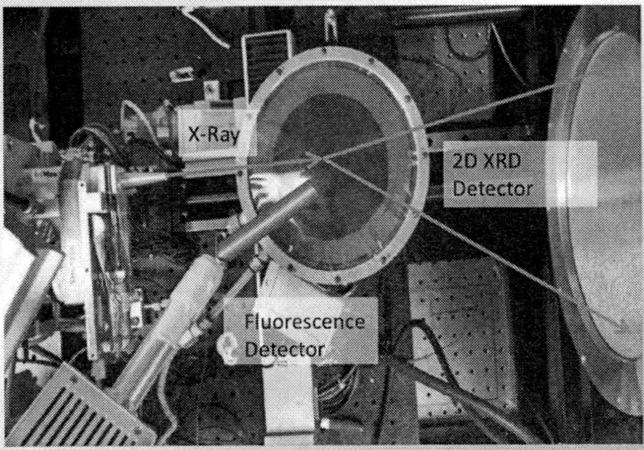

Oxide Growth Study through *in-situ* XRD

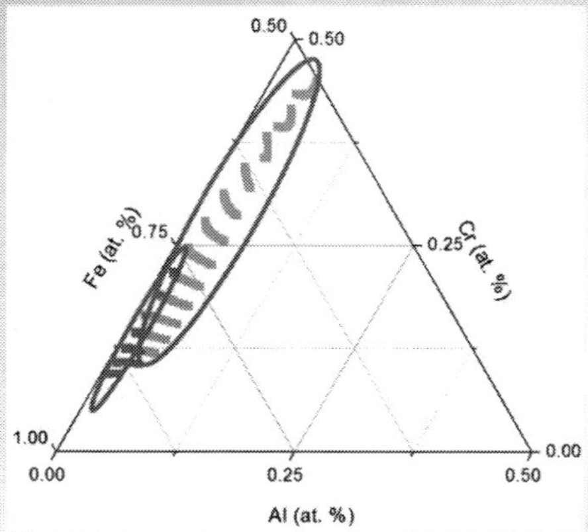

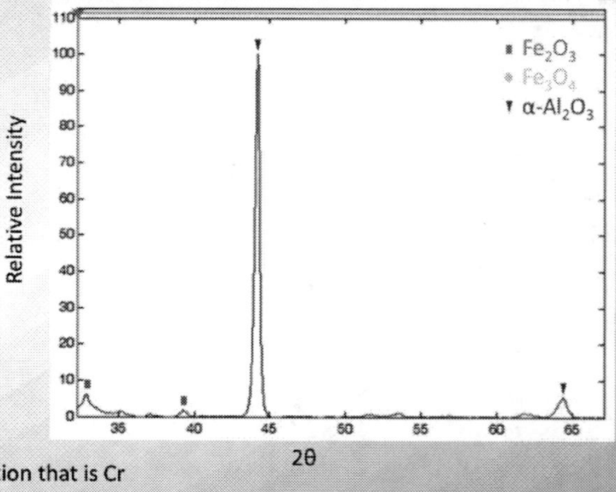

- Iron oxides are suppressed when the composition that is Cr rich (small amount are observed)

Comparison of Trends Between Low and High Temp Oxidation

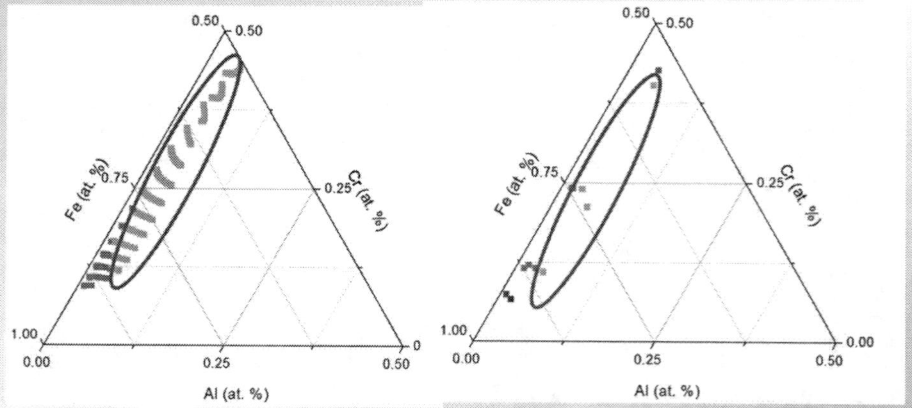

- XRD and Raman matched up where the more Al and Cr rich region are a lot easier to form α-Al_2O_3

Summary

- Probed the temporal effect of temperature on the metallurgic and oxide phase transitions
- A method for the **initial high-throughput and non-destructive screening** of FeCrAl thin film was developed
- **Cr promotes** α-Al_2O_3 growth on the surface of the samples to form a protective layer preventing further development of other oxides
  - This indicates that Cr rich FeCrAl are more favored to operate under high temperature conditions
- Oxidation studies indicate that severe oxide growth was not observed until 400°C
- The methodology can be applied to more complicated systems for high temperature oxidation resistant materials.
- Detailed studies of **oxidation and phase transitions of single compositions at different temperatures** will refine the results

15

Acknowledgement

- **Dr. Jason Hattrick-Simpers**
- **Jonathan Kenneth Bunn**
- **SmartState Center for Strategic Approaches to the Generation of Electricity**
- **USC Honors College (Science Undergraduate Research Fellowship Program)**
- **University of South Carolina**
  - **College of Engineering and Computing Science**

This project was funded in part by the South Carolina Honors College

16

UNIVERSITY OF SOUTH CAROLINA
College of Engineering and Computing

SAGE
Power through Engineering

Thank you for your Attention

Are there any questions?

17

Crossing Borders and Communities for Healthy Water

Joshua Gomez, Meng Zhou, Dr. Shuguang Deng

Chemical and Materials Engineering Department, New Mexico State University, Las Cruces, New Mexico, 88003

In Partnership with:

NM Water Resources Research Institute

NMSU Chemical Engineering Department

The New Mexico Alliance for Minority Participation

New Mexico Environment Department

Border Partners (Non-profit)

EPA 2020

National Science Foundation

New Mexico State University NM STATE

Puerto Palomas, Chihuahua, MX

- Located just south of Columbus, NM.
- Population of 4,688.
- Arsenic contamination (0.030-0.1 mg/L).
  - Horrible Carcinogen.
- Fluoride contamination (5-40 mg/L).
  - Bone Fluorosis and neurological damage.
- Concentration depends entirely on well water.
  - Naturally occurring concentrations of arsenic and fluoride exceed the permissible limits of Mexico and the U.S.

New Mexico State University NM STATE

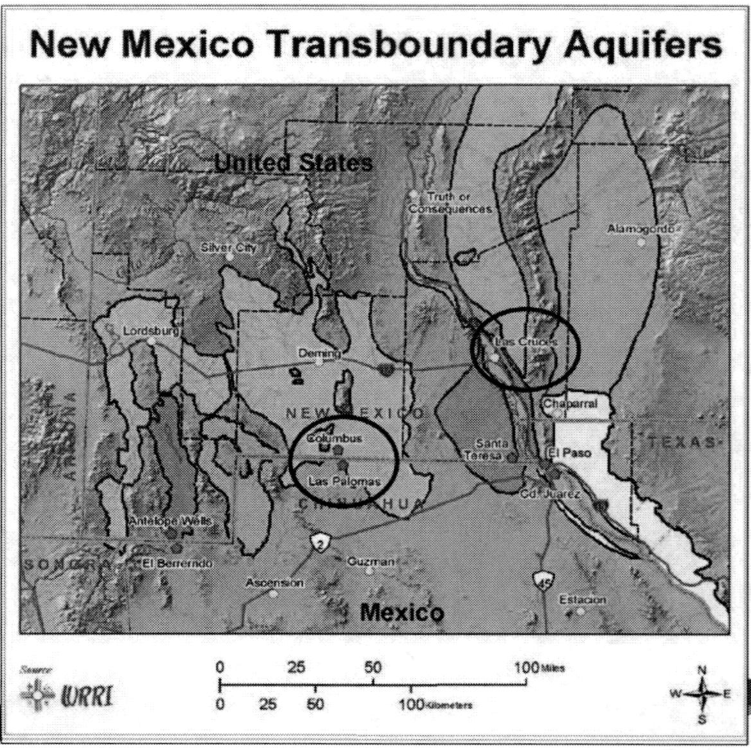

New Mexico State University

Proposed Solution

- Homemade water filter
  - Low cost <$15.00
  - Easy to assemble
  - Removes arsenic
  - Removes fluoride
  - Utilizing activated carbon (AC)
  - Utilizing activated alumina(AA)
    - Iron coated
  - Improves the taste

New Mexico State University

Filter Designs

- 1 filter of the final design costs $5.00 to make.
- The Activated Alumina costs $1/kg.
- The Activated carbon used cost $35/lb.

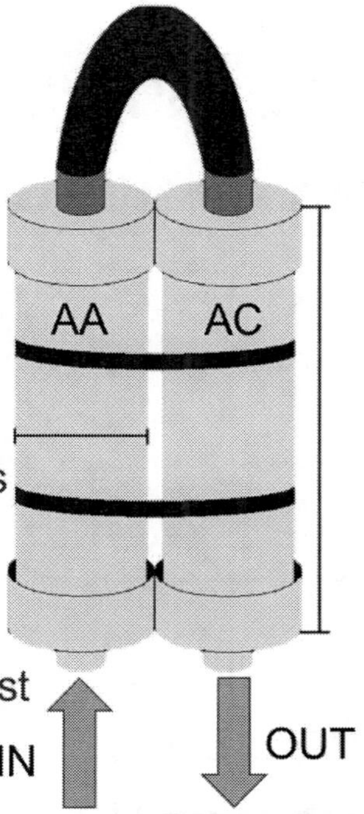

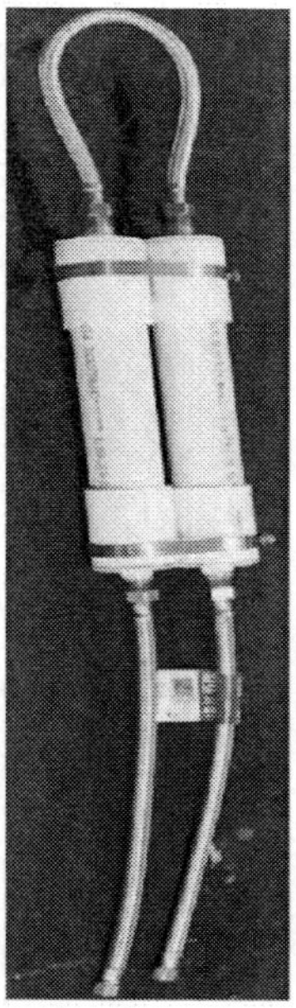

New Mexico State University

Our Approach

NMSU demonstrated to Palomas science teachers how to assemble a simple water filter.

Science teachers Juan Carlos Huerta and Joel Carreon pose with the filter they assembled for installation at the Palomas primary school.

New Mexico State University

What's Working

Hall Environmental Laboratory, Albuquerque

CONTAMINANT	RESULTS	STANDARDS					
		U.S.	Exceeded	%	Mexico	Exceeded	%
Fluoride							
Unfiltered	4.90 mg/L	4.00 mg/L	Y	22.5	1.50 mg/L	Y	226.7
Filtered	0.240 mg/L	4.00 mg/L	N	-	1.50 mg/L	N	-
Arsenic							
Unfiltered	0.030 mg/L	0.01 mg/L	Y	200.0	0.02 mg/L	Y	20.0
Filtered	ND	0.01 mg/L	N	-	0.02 mg/L	N	-

ND = Not Detected

New Mexico State University

Optimization of Algae Derived Activated Carbon (AC)

- Low cost alternative to commercial activated carbon.
- More effective adsorbent than commercial carbon.
- Chlorella algae is used as a carbon source.
- Potassium Hydroxide and Iron (II) Chloride
  - Activating agent.
- Carbonization times of 500° C, 600° C, and 700° C.
- Shaker experiments to determine most effective adsorbent for Arsenic(III).

New Mexico State University

Carbon Experimental data

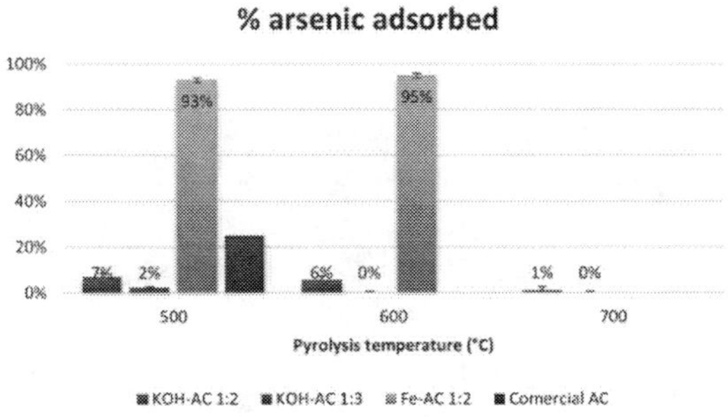

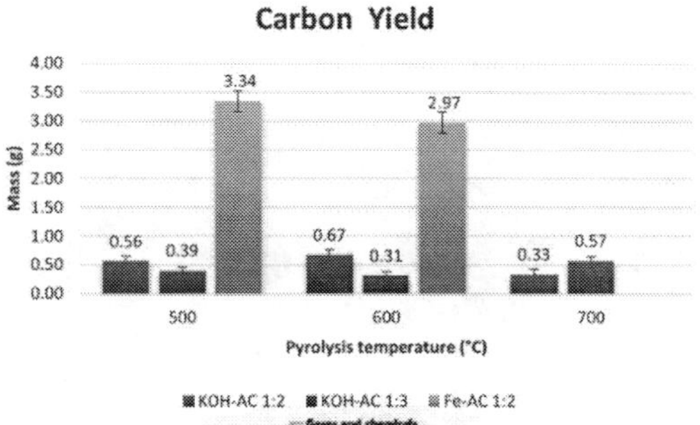

adsorption capacity (mg/g)			
	500°C	600°C	700°C
KOH-AC 1:2	0.082	0.067	0.010
KOH-AC 1:3	0.024	0.000	0.000
Fe-AC 1:2	1.143	1.168	-
Comercial AC	0.275	-	-

$$capacity = \frac{(InitialConcentration - FinalConcentration)Volume}{\text{Mass of adsorbent}}$$

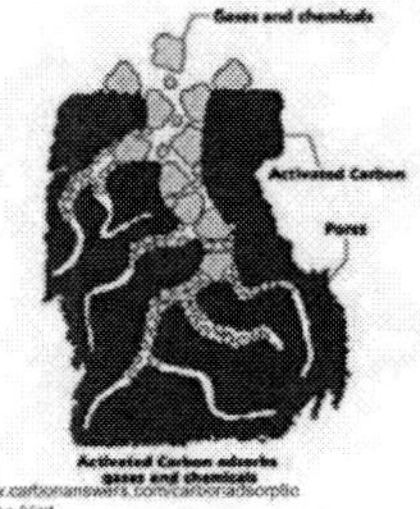

Fasci, M.

New Mexico State University

X-Ray Difraction Pattern

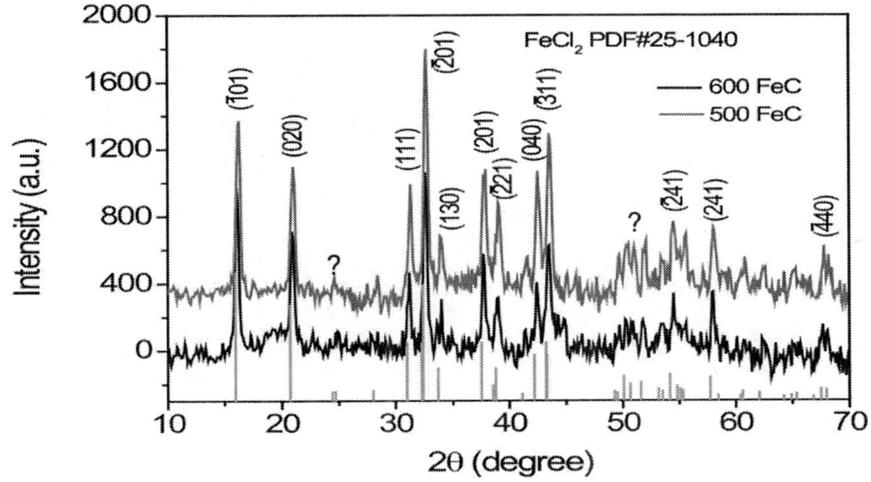

New Mexico State University

Scanning Electron Microscope Imaging

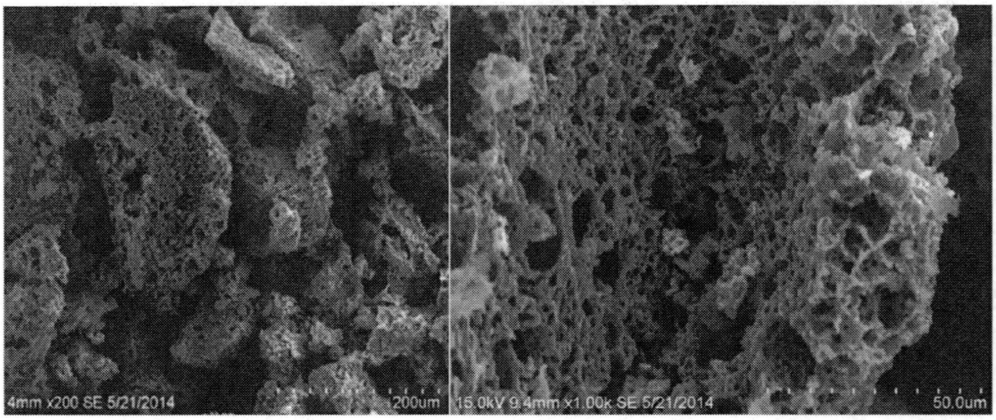

New Mexico State University

Unwashed FeCl(II) Activated Carbon Textural Properties

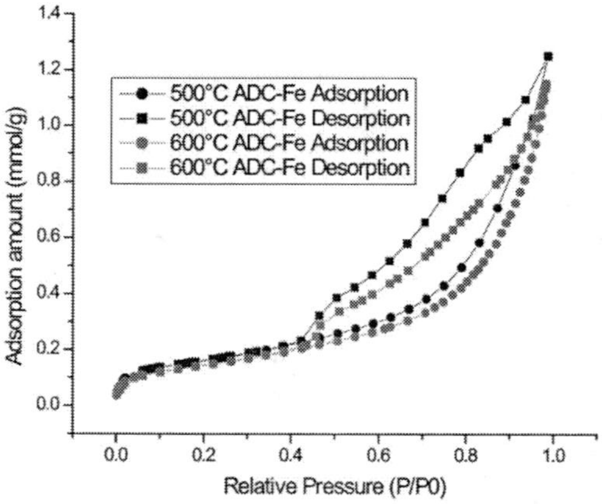

N_2 adsorption and desorption isothermal @ 77K for unwashed samples.

	500 ° C FeCl2/ADC	600 ° C FeCl2/ADC
BET surface area (m^2/g)	12.9893	11.6336
Average pore size (nm)	8.6408	6.5769
Pore volume (cm^3/g)	0.0394	0.0476

New Mexico State University

Washed FeCl(II) Activated Carbon Textural Properties

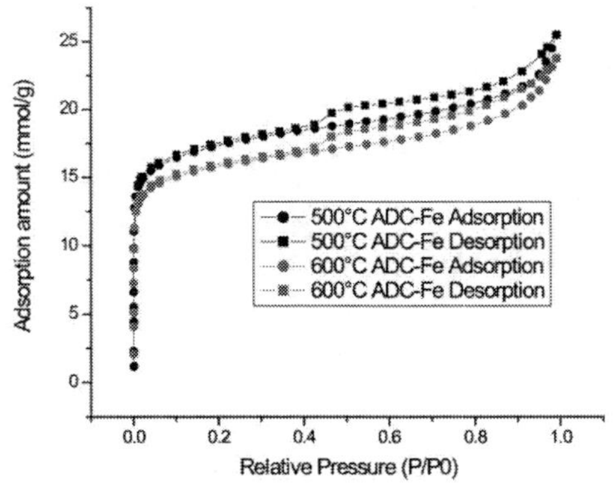

	500 ° C ADC (after washing)	600 ° C ADC (after washing)
BET surface area (m^2/g)	1326.0368	1208.1831
Average pore size (nm)	7.8704	7.9340
Pore volume (cm^3/g)	0.3362	0.3305

N2 adsorption and desorption isothermal @ 77K for washed samples.

Shaker Experiments

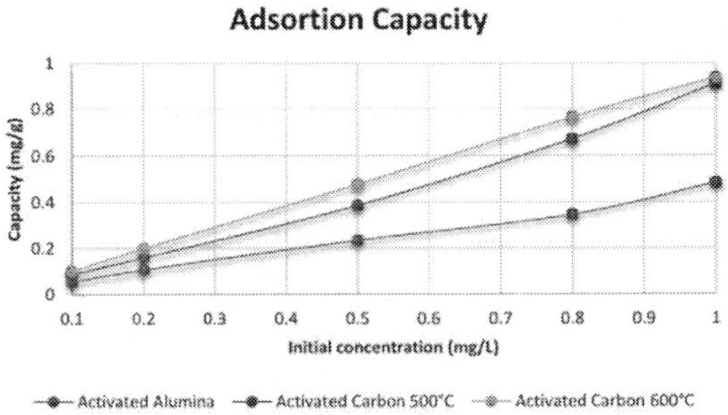

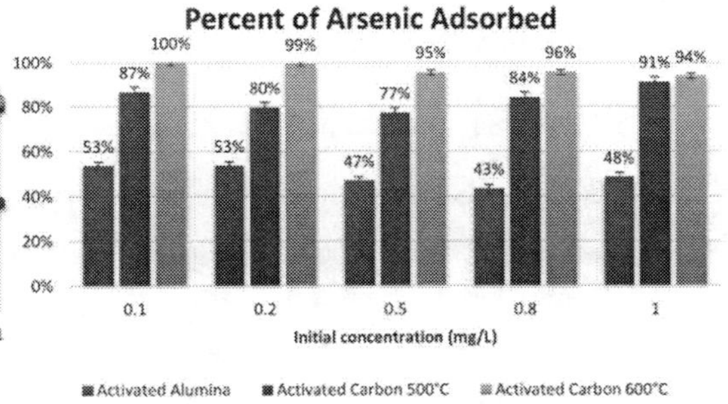

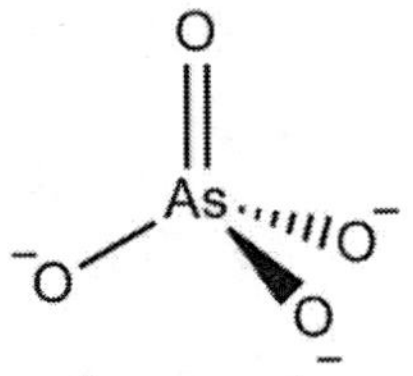

Arsenate

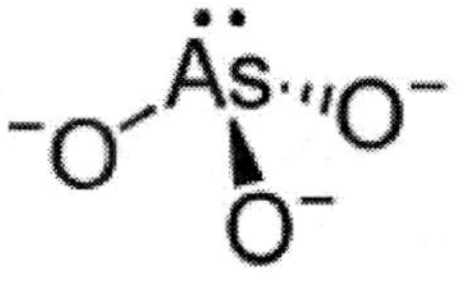

Arsenite

Conclusions

- Filters are still being made in Palomas by the teachers today and will continue to be made.
- The results of the shaker experiments and textural property data shows that the carbon activated with Iron (II) Chloride is a viable replacement.

New Mexico State University NM STATE

Next Steps

- Further Development of the filter.
- Reach out to other communities.
- Advance research with optimized carbon.

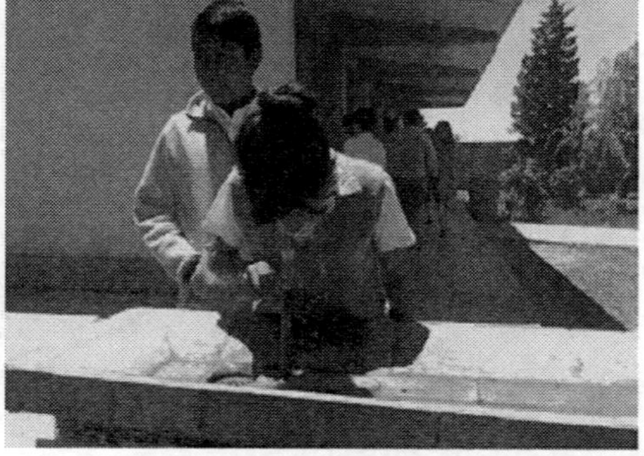

New Mexico State University NM STATE

Acknowledgments

- Dr. Shuguang Deng, Chemical & Materials Engineering dept.
- Meng Zhou, Chemical & Materials Engineering dept.
- Dr. David A. Rockstraw, Chemical & Materials Engineering dept.
- Erin Ward, Water Resources Research Institute.
- Mark Chidester, Chemical & Materials Engineering dept.
- Peter Edmunds, Border Partners.

NEW MEXICO WATER RESOURCES RESEARCH INSTITUTE

NEW MEXICO AMP

NSF HRD # 1305011

Deng's Renewable Energy and Advanced Materials

New Mexico State University

Questions?

Email: jgomez38@nmsu.edu

New Mexico State University

394186 Ferroelectric BTO on Si (001) for High-Efficiency Solar Cell Heterostructures

Abstract:

Ferroelectric oxides such as barium titanate (BTO) have recently become a topic of interest in relation to thin-film solar cell production due to the material's stability at a wide range of temperatures, its spontaneous polarization, and the potentially low-cost of production. Integration of BTO on silicon could enable the creation of highly efficient solar cells, as BTO's ferroelectric properties encourage the detachment of photo excited electrons. We have previously demonstrated successful integration of BTO on silicon carbide (SiC) substrates, using a magnesium oxide (MgO) heteroepitaxial layer to enable single-crystal orientation of the BTO film deposited on SiC.[1] BTO has also been successfully grown on silicon substrates, but with the use of strontium titanate (STO) and MgO buffer layers to resist formation of SiO_2.[2] Integration of BTO on Si with a single MgO interface layer would decrease interfacial losses in a final solar cell device. Based on previous work with SiC, we believe a single interface can be possible by controlling the surface interactions during film growth.

In this research, MgO is grown on Si (001) substrates in an ultrahigh vacuum (UHV) environment using a number of growth methods including molecular beam epitaxy (MBE) and sputtering. In comparing differences in surface structure using reflection high-energy electron diffraction (RHEED) and surface chemistry using X-ray photoelectron spectroscopy (XPS), we can understand surface interactions between Si, Mg, and O. By understanding the relationship between atomic level interactions and ultimate film characteristics, we can engineer the most effective process for BTO integration on Si. The substrates are cleaned using wet chemicals to create a hydrogen-terminated surface that resists contamination from laboratory exposure. XPS and RHEED are used to verify the lack of contamination on the surface and proper crystallographic orientation. Contaminant levels of 7.61 % oxygen (0.15 O/Si ratio) and 18.0% carbon (0.31 C/Si ratio) are the most successful results that have thus far been attained. XPS and RHEED will serve to confirm proper growth of MgO (001) on Si and determine thickness of the MgO film. BTO is then grown on the correctly-oriented MgO film using MBE, as has already been demonstrated on MgO/SiC substrates.[1] This presentation will discuss cleaning and growth conditions for successful production of high quality BTO on Si using the single MgO interface layer.

393873 Natural Gas, the Bridge Fuel

Abstract:

Medium and heavy duty trucks and buses produce almost 18 percent of all carbon emissions from transportation. Alternative energy sources need to be used on a larger scale to reduce these emissions, as well as to reduce dependence on foreign oil. Natural gas is the future of fuel in large vehicles. This paper tackles different issues surrounding the usage of natural gas as the main fuel for natural such as public awareness, public policy, existing infrastructure, consumer choice and controversies surrounding hydraulic fracturing. Current pilot programs, in which public city buses are using natural gas as the fuel of their choice, will be analyzed to determine the best way to implement these programs in other areas of the United States as well.

397811 The Federal Role in Fostering an Innovative U.S. Energy Ecosystem

Abstract:

Although renewable sources of energy – solar, wind, hydro, and geothermal – have been around for over one hundred years, they only generate 12.5 percent of electricity in the United States today. Renewable technology has advanced in cost, reliability, performance, and efficiency, but not enough to allow renewables to be a large contributor to electricity generation and widely used in the U.S. Renewables need continued innovation and progression to become more efficient, more reliable, and more cost competitive with fossil fuels, as well as expanding the possibilities of their use. Increased R&D funding for renewable sources is the key to advancing the technology and giving renewables the tools it needs to provide a larger portion of electricity in the U.S.

With renewables generating a larger portion of electricity, the U.S. can reach environmental sustainability, energy security, and a new economic opportunity. In reference to reaching environmental sustainability, renewables do not emit carbon and greenhouse gases, which is important for a cleaner future and mitigating climate change. Because the "fuel" for renewables includes the sun and wind, electricity generation from these sources does not depend on obtaining fossil fuels inside or outside of our country. They also extend the lifetime of these resources to be used for base load power, providing energy security to the U.S. Finally, if the U.S. is able to quickly and effectively develop competitive renewable technologies and deploy them into the electricity sector, it can provide competitive renewable technologies to the global community, boosting the economy and maintaining world leadership.

Although renewables have the capability to provide many benefits for the U.S., they face obstacles in receiving the funding they need to be widely used. These obstacles include climate change skeptics, the fossil fuel industry, the polarized political environment, and having to cut funding for other areas such as education and defense in order to allocate more money to R&D. The demand for renewables has also been relatively weak because electricity generated by renewables is no different than electricity generated by fossil fuels when the consumer uses it. The main way for renewables to be competitive with fossil fuels is through the cost and since wind is the only renewable that is cost competitive with fossil fuels, a strong customer demand has not been present to spur large private sector investment in renewable technologies.

In order for renewables to accomplish necessary advancements and contribute a larger portion of electricity to the U.S., financing to R&D for the technologies needs to be increased through government funding, capital generated through a carbon tax, public-private partnerships, and incentives for investment in the technology and its R&D. A commitment to innovating renewables in order to reach environmental sustainability, energy security, and a new economic opportunity is necessary from governmental officials to make the tough decisions advocating for widespread use of renewables in the U.S.

397856 A Biowall for Improving Indoor Air Quality

Abstract:

The Biowall is a living plant wall that filters air in the HVAC system of a home or small commercial building. This new technology improves indoor air quality be removing Volatile Organic Compounds and has the potential for energy savings by reducing the need for heating and cooling large amounts of outside air in a building. The presentation provides an overview of biofiltration for buildings and discusses the new research being conducted in this emerging field.

376510 Artificial Photosystem I and II: Highly Selective Solar Fuels and Tandem Photocatalysis

Abstract:

Artificial photosynthesis, or generation of solar fuels from CO_2 /H_2 O, can provide an important alternative for rising CO_2 emission and renewable energy generation. In our recent work, composite photocatalysts (CPCs) made from widebandgap nanotubes and different QDs were used to mimic Photosystem II (PS680) and I (PS700), respectively. By tuning the redox potentials using the size, composition and energy band alignment of QDs, we demonstrate highly selective (> 90%)and efficient production of ethane, ethanol and acetaldehyde as solar fuels with different wavelengths of light. We also show that this selectivity is a result of precise energy band alignments (using cationic/anionic doping of nanotubes, QD size etc.), confirmed using measurements of electronic density of states, and alignment of higher redox potentials with hot-carriers can also lead to hot-carrier photocatalysis. This wavelength-selective CPCs can have important implications for inexpensive production of solar fuels including alkanes, alcohols, aldehydes and hydrogen, and making tandem structures (red, green, blue) with three CPCs, allowing almost full visible spectrum (410   730nm) utilization with different fuels produced simultaneously.

385082 Transport and Dynamics of Ionic Species in Block Copolymer Electrolytes for Solid-State Lithium Batteries

Abstract:

Solvent-free block copolymer electrolytes are promising materials for solid-state lithium batteries, as their tunable compositions and nanoscale morphologies enable simultaneous control of their ionic conductivities and mechanical properties. However, quantifying and distinguishing the transport and dynamics of the different ionic species can be challenging, which are linked to the complex hierarchy of motional processes of the polymer chains. We measured ion transport properties and dynamics with solid-state nuclear magnetic resonance (NMR) spectroscopy in different poly(ethylene oxide) (PEO)-based polymer electrolytes containing lithium bis(trifluoromethane)sulfonomide (LiTFSI), including neutral and anionic triblock copolymers and their homopolymer analogue. The diffusion coefficients of the Li^+ and $TFSI^-$ ions, apparent diffusional activation energies, ion transport numbers, and tortuosity factors were established quantitatively by ^7Li and ^{19}F pulsed-field-gradient (PFG) NMR measurements, which are compared with results from electrical conductivity measurements. Characteristic correlation times of motion were extracted from ^7Li and ^{19}F NMR relaxometry, yielding insights into ion dynamics over nanosecond timescales. *In situ* ^7Li NMR imaging measurements of symmetric lithium metal-polymer electrolyte cells yield insight into the growth and inhibition of lithium dendritic structures through the different polymer electrolytes. The results are expected to aid the rational design of new block copolymer electrolytes with controllable ion transport and mechanical properties.

Investigating the chemical composition and structure of the cathode-electrolyte interface layer in lithium manganese oxide batteries using reactive force field (ReaxFF) based molecular dynamics

Lithium ion based batteries have great potential to revolutionize our transportation sector. Among the various lithium metal oxides that have been used, or proposed for use, as cathode active materials, lithium manganese oxide ($LiMn_2O_4$) is the primary cathode material being exploited for automotive applications, on account of its low cost, low toxicity, and simple preparation process. There are, however, well documented problems with $LiMn_2O_4$, the most critical of which is capacity fade due to active material dissolution. The dissolution of divalent manganese from the surface of the cathode into the electrolyte leads to a decrease in the effective amount of cathode material available for further lithium intercalation. The deposition of the dissolved manganese ions on the anode also decreases the Li content of the anode thereby decreasing the overall cell capacity even more. Experimental observation suggests that the manganese content in the electrolyte increases as the cell ages, making the loss of cathode material the primary reason for the reduced cell life for a $LiMn_2O_4$ battery.

The key to improving the performance of $LiMn_2O_4$ as a cathode material is to understand the reactions occurring at the surface of the cathode which can then be used to increase the cathode stability and prevent the dissolution of manganese. The presence of divalent manganese ions in the electrolyte has been detected by experiments, but its speciation and related reaction mechanisms have not yet been fully understood. Tunneling electron microscopy imaging techniques have shown the presence of a layer of electrolyte decomposition products at the cathode surface, but no further composition analysis has been possible due to its thinness and delicate structure. In order to bridge this knowledge gap, this work applies molecular modeling and density functional theory (DFT) methods to investigate the reactions occurring at the $LiMn_2O_4$ cathode surface and the mechanisms of manganese dissolution. Specifically, a reactive force field (ReaxFF) has been developed to conduct molecular dynamics (MD) simulations of these electrochemical cells. Reactive force fields can be used to simulate the formation and dissociation of chemical bonds during a molecular dynamics simulation, which will help disclose the extremely dynamic surface reaction chemistry of the cathode.

The results for ReaxFF MD simulations of the $LiMn_2O_4$ cathode-electrolyte interface layer will be presented. The chemical composition and structure of this layer will be compared with experimental observations. The similarities and differences between the cathode interface layer and well-studied anode interface layer will be discussed.

In situ Formed Sulfur/Carbon Composites for Advanced Li-Sulfur and Na-Sulfur Batteries

Chao Luo, Jingjing Wang and Chunsheng Wang

Abstract: Li-sulfur and Na-sulfur batteries are considered as promising energy storage devices for emerging electric vehicles and smart grids due to the low cost and high capacity of sulfur. However, two major challenges limit practical application of Li-sulfur and Na-sulfur batteries: (1) The low sulfur utilization due to insulating nature of sulfur and sulfides; (2) Severe shuttle reaction due to the dissolution of polysulfide intermediates. Although LiTFSI-DOL/DME electrolyte with high polysulfides solubility is widely used to enhance the utilization of sulfur, the severe shuttle reaction significantly reduces the Coulombic efficiency for Li-sulfur and Na-sulfur batteries. One of the most effective methods in suppressing the shuttle reaction is constraining polysulfides in porous carbon matrix. In this work, sulfur/carbon composites were synthesized by annealing the mixture of sulfur and perylene-3, 4, 9, 10-tetracarboxylic dianhydride in a sealed glass tube. The *in situ* formed sulfur/carbon composites exhibit superior electrochemical performance in commercial carbonate based electrolyte, $LiPF_6$-EC/DEC. A reversible capacity of 520 mAh g^{-1} is maintained for over 1000 cycles in the cutoff window from 1.0 V to 3.0 V. When the cell is fully discharged to 0.6 V in the first five cycles, and then cycled in the cutoff window from 1.0 V to 3.0 V, the reversible capacity is remarkably enhanced to 820 mAh g^{-1} for over 600 cycles. The large capacity increase is attributed to the fully lithiation of sulfur. The strong interaction between sulfur and carbon matrix due to the *in situ* formation lowers the utilization of sulfur, but the fully discharge to 0.6 V in the first five cycles alleviates the strong interaction between sulfur and carbon matrix to allow more sulfur to react with lithium ions, resulting in increase of capacity. The excellent electrochemical performance is also achieved for Na-sulfur batteries that a reversible capacity of 400 mAh g^{-1} is maintained for 150 cycles. Therefore, the *in situ* formed Sulfur/Carbon Composites are promising cathode materials for Li-sulfur and Na-sulfur batteries.

379845 Formation of Large Polysulfide Complex during Lithium-Sulfur Battery Discharge

Abstract:

Sulfur cathodes have much larger capacities than the transition-metal-oxide cathodes used in commercial lithium-ion batteries, but suffer from unsatisfactory capacity retention and long-term cyclability. Capacity degradation originates from soluble lithium polysulfides gradually diffusing into the electrolyte. Understanding the formation and dynamics of soluble polysulfides during the discharging process at the atomic level remains elusive, which limits further development of lithium-sulfur (Li/S) batteries. Here we report first-principles molecular dynamics simulations and density functional calculations, through which the discharging products of Li/S batteries were studied. We find that instead of simple Li_2S_n ($1 \leq n \leq 8$) clusters generated from single cyclooctasulfur (S_8) rings, large Li/S clusters form by collectively coupling several different rings to minimize the total energy. While some small clusters bind to the electrolyte molecules, such as DOL and DME, and diffuse out, the remainders conglomerate into large Li/S clusters. At high lithium concentration, a Li-S network forms at the sulfur surfaces. Our results explain the formation of insoluble Li-S complex at the end of the discharging process and the slow dynamics of phase transformation from Li_2S_2 to Li_2S. In addition, we show that oxygen impurities in graphene, particularly oxygen atoms bonded to vacancies and edges, may stabilize the lithium polysulfides that may otherwise diffuse into the electrolyte.

Support was provided by the Petroleum Institute, Abu Dhabi, and the McMinn Endowment at Vanderbilt University. DFT calculations were performed at the DoD AFRL.

Section: Nanomaterial for energy

Title: An Advanced MoS_2/Carbon Anode for High Performance Sodium-Ion Batteries

Authors: Jingjing Wang, Chao Luo, Alex Langrock, Alice C. Mignerey, Chunsheng Wang

The concerns on the availability and distribution of lithium resource in the earth's crust render the necessity to develop new battery chemistries. Among them, sodium-ion battery is unambiguously the primary candidate due to low cost, large resource availability, and similar insertion chemistry with lithium ions. However, the larger ion size of Na^+ than Li^+ induces more complicated and difficult insertion chemistry, and fast capacity decay is still a critical issue for anode materials, which impedes the widespread application of Na-ion batteries. Therefore, it is of great importance to design and fabricate advanced anode materials with high capacity and long cycle life. Molybdenum disulfide (MoS_2) is a promising anode for high performance sodium-ion batteries due to its high specific capacity, abundance, and low cost. However, poor cycling stability and low rate capability induced by large volume expansion during sodiation are the main challenges for using MoS_2 in Na-ion batteries. In this study, molybdenum disulfide/carbon (MoS_2/C) nanospheres with flower-like morphology were fabricated using hydrothermal method for Na-ion battery anodes. The flower-like MoS_2/C nanospheres deliver a reversible capacity of 520 mAh/g at 0.1 C and maintain at 400 mAh/g for 300 cycles at a high current density of 1 C, demonstrating the best cycling performance of MoS_2 for Na-ion batteries up to date. The high capacity of MoS_2/C nanospheres is attributed to the short ion and electron diffusion pathway and large electrolyte/electrode contact area, which enable fast charge transfer and low concentration polarization. The stable cycling performance and high coulombic efficiency (~100%) of MoS_2/C nanospheres are ascribed to (1) highly reversible conversion reaction of MoS_2 during sodiation/desodiation as evidenced by *ex-situ* X-ray diffraction (XRD) and (2) the formation of a stable solid electrolyte interface (SEI) layer in fluoroethylene carbonate (FEC) based electrolyte as demonstrated by fourier transform infrared spectroscopy (FTIR) measurements. To our best knowledge, the reversible conversion reaction mechanism of MoS_2 during sodiation/desodiation has never been reported.

Modeling Hydrogen Storage of Metal-Organic Frameworks

Yasemin Basdogan[1], and Seda Keskin[1,2*]

[1]*Department of Chemical and Biological Engineering, Koc University, Rumelifeneri Yolu, Sariyer, 34450, Istanbul, Turkey*

[2]*KÜTEM (Koc University TUPRAS Energy Center), Koc University Rumelifeneri Yolu, Sariyer, 34450, Istanbul, Turkey*

Developing sustainable transportation technologies is an increasing topic and has highlighted the need for high-density hydrogen storage. Hydrogen usage for transportation is one of the most promising solutions to today's transportation problem. Considering hydrogen as a non-petroleum energy carrier has several benefits compared to petroleum carriers. Hydrogen has high energy content, gives exhaust products rather than greenhouse gasses and can be derived from a variety of primary energy sources. There is only one obstacle in using hydrogen as a non-petroleum energy carrier that is the storage of low-density gaseous hydrogen. There are several methods used such as liquefaction and physical compression, however none of these methods are efficient enough to be used. Physical adsorption of hydrogen on large surface areas is one of the most appealing options. New classes of microporous materials called metal organic frameworks (MOFs) are excellent candidates for reversible adsorption of hydrogen considering their extremely high surface areas and tunable structures. Adsorption takes place on the surface of the MOF by means of physical adsorption. This kind of adsorption behavior is called physisorption, which makes it possible to observe totally reversible gas uptake and release behavior. For all the above reasons MOFs are attractive materials for hydrogen storage (1).

We initially investigated the factors that affect the hydrogen storage in MOFs: heat of adsorption, surface area and pore volume. Our main goal was to achieve a model that can predict the hydrogen uptake of a MOF by using these three key factors. First, we began our research by validating the simulation results that are obtained from different research groups. We put together a table that summarizes the most current hydrogen simulation studies. In this table, we included the type of simulation performed by the researcher, type of the force field and the hydrogen model that they used in their simulations. We later compared these simulation results with the experimental uptakes and observed that the simulation results are mostly in good agreement with the experimental results. We realized that almost every researcher used Grand Canonical Monte Carlo simulation method, however each researcher used a different force field (UFF, Dreiding or a new force field that they have calculated from ab initio calculations) and different hydrogen parameters (single site Lennard Jones sphere, two site Lennard Jones sphere or Morse potential). We included the specific temperature and pressure of the simulations in order to compare the results of simulations with the experimental results. After demonstrating the accuracy of the simulation results, we performed simulations for a large group of MOFs at a wide pressure range at 77 K.

To be accurate, we collected experimental data for 20 of MOFs and crosschecked them with our simulation results. We obtained a regression coefficient (R^2) of 0.80 which means that the simulation results are in fairly good agreement with experimental results considering the experimental errors. We investigated the relationship between hydrogen uptake and the three factors listed above. We calculated an $R^2 \approx 0.92$ for a linear relationship between hydrogen uptake and surface area, an $R^2 \approx 0.96$ between hydrogen uptake and pore volume. These calculations showed that we should focus on an equation that linearly depends on surface area and pore volume. The R^2 value between heat of adsorption and hydrogen uptake was about 0.56 which suggests that, heat of adsorption does not play a dominant role in predicting hydrogen adsorption or it does not have a linear relationship with the hydrogen uptake. We calculated the coefficients for our equations by using a Matlab code based on minimization of the standard deviation.

After validating our simulation results we focused on four pressures (1, 10, 35, 100 Bar). We used the algebraic formula (0.0021×Surface Area+0.1×Pore Volume) that we found from literature for 1 Bar (2). With the initial formula that we have used we obtained an $R^2 \approx$ 0.33. We tried to change the coefficients for surface area and pore volume, however the results did not improve. We included the heat of adsorption in the linear relationship and added a constant to the equation which improved R^2 to $\approx$0.39. As the pressure increases we obtain better results. For 10 Bar, a linear equation containing surface area and pore volume gave an $R^2 \approx 0.797$. When we added the heat of adsorption to the linear equation, we observed a slight increase at $R^2 \approx 0.799$. For 35 Bar, both equations (one containing only surface area and pore volume other additionally containing the heat of adsorption) gave the same $R^2 \approx 0.945$. When we increased the pressure to 100 Bar, we obtained our best-fit model. We succeeded to achieve an $R^2 \approx 0.982$, again yielding exactly the same result with two different linear models. By the end of this research we realized that heat of adsorption does not improve accuracy in modeling hydrogen uptake. Heat of adsorption does not dominate the hydrogen uptake in a significant manner, which proves our beginning hypothesis. On the other hand surface area and pore volume played an important role predicting the hydrogen uptake without the need of simulations or experiments.

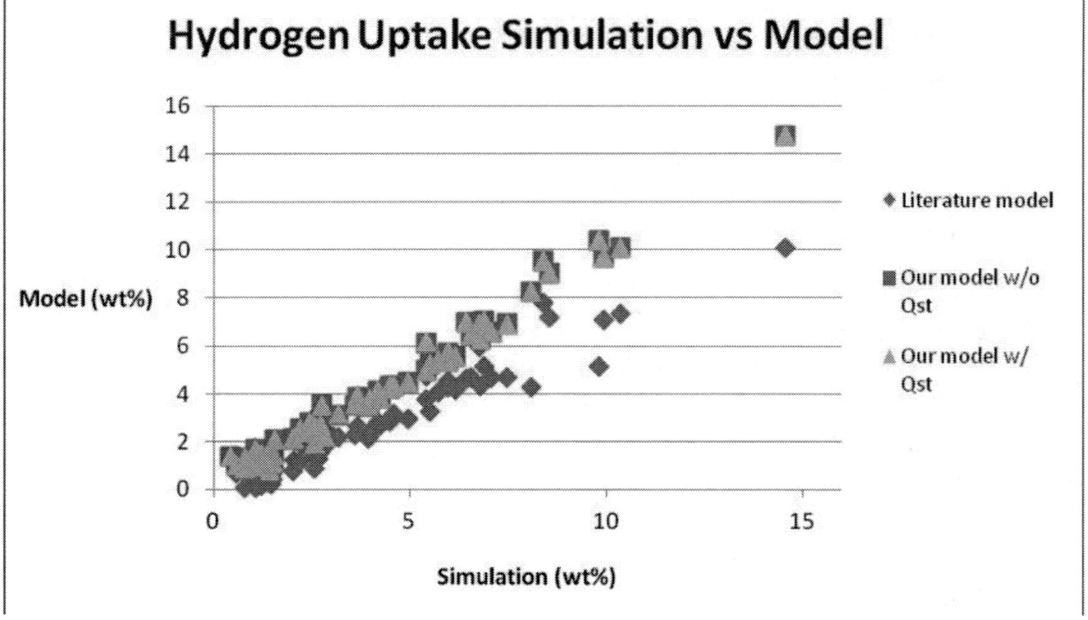

Figure 1: Hydrogen adsorption calculated from model versus the hydrogen adsorption obtained from the simulations at 77 K and 100 Bar. (Qst: Heat of adsorption)

In conclusion, we used eight equations to predict the hydrogen adsorption of MOFs. Our accuracy improved as the pressure increases and we succeeded to show that heat of adsorption does not play a crucial role in predicting hydrogen uptake. We tried to generate these models because we aimed to predict hydrogen adsorption of MOFs just by simple calculations. Accurate models can be used as a reverse engineering tool to find the optimum material that can store hydrogen to reach the U.S. Department of Energy (DOE) targets. DOE in cooperation with the U.S Council for Automotive Research (USCAR) established several evolving targets for hydrogen storage systems. The most upcoming target is for 2015, which is 5.5% (system gravimetric capacity). DOE and USCAR will support the materials reaching this target. Our future aim is to enter this target into the models that we have found, and calculate the surface area or the pore volume by leaving the other variable constant and reaching for the best material that can store hydrogen.

1. Theoretical Limits of Hydrogen Storage in Metal–Organic Frameworks: Opportunities and Trade-Offs J. Goldsmith, A. G. Wong-Foy, M. J. Cafarella and D. J. Siegel *Chem. Mater.***2013** *25* (16), 3373.
2. The current status of hydrogen storage in metal–organic frameworks-updated J. Sculley, D. Yuan and H.-C. Zhou, Energy Environ. Sci., 2011, 4, 2721, Chem. Rev., **2012**, *112*, 724.

Experimental Investigation of Hydrogen Storage in Silicon Carbide Nanotubes

SeyedHamed Barghi, Theodore T. Tsotsis, and Muhammad Sahimi[†1]

Mork Family Department of Chemical Engineering & Materials Science, University of Southern California, Los Angeles, California 90089-1211, United States

Abstract

The experimental results of hydrogen storage in silicon-carbide nanotubes (SiCNTs) are reported in this work for the first time. The SiCNTs are produced by gas-solid reaction between silicon monoxide (SiO) vapor and carbon nanotube (CNT) precursors at temperatures between 1200 °C-1500 °C, which is the temperature range used in most of the previous studies for SiCNTs synthesis [1-6]. The SiO vapor was produced by a solid-solid reaction between silicon (Si) and silicon dioxide (SiO_2) during the SiCNTs synthesis process. A purification process based on sodium hydroxide was then carried out on the as-synthesized SiCNTs. The purpose of the purification process was to remove the side products of the synthesis reactions from the surface of the SiCNTs. Using a novel gravimetric method coupled with a residual gas analyzer, the hydrogen uptake behavior of the as-synthesized SiCNTs and also the purified SiCNTs was then studied at room temperature and at pressures of up to 100 bar.

According to the experimental results, the hydrogen storage capacity of the as-synthesized SiCNTs is comparable to that of the CNT precursors. On the other hand, the purified SiCNTs show a 50% higher hydrogen storage capacity. Moreover, the hydrogen uptake occurs five time faster in the SiCNTs compared to the CNTs. The hydrogen desorption is also completely reversible in the SiCNTs, but this is not the case for the CNTs [7]. SiCNTs present a superior alternative to CNTs, not only for hydrogen storage, but potentially for other applications at high temperatures, often used during hydrogen production, because of (i) the excellent mechanical

[†] Corresponding author. Tel.: +1 (213)740-2064 E-mail address: moe@usc.edu

properties of the SiCNTs; (ii) their resistance to harsh corrosive conditions in the presence of high-temperature steam.

Among the as-synthesized SiCNTs, the ones synthesized at 1200 °C showed the highest hydrogen uptake capacity, whereas those that were prepared at 1500 °C possessed the lowest capacity. This difference in hydrogen storage capacity is mostly due to the differences in their SiO_2 impurity content (The presence of the amorphous SiO_2 impurity was confirmed by FTIR analysis). The evaporation of the unreacted SiO_2 is the source of this impurity. Since the vapor pressure of SiO_2 is lower at lower temperatures, the SiCNTs prepared at the lower temperatures contain less impurity as well. The SiO_2 impurity has a negative effect on the hydrogen storage capacity of the SiCNTs because:

(i) It acts as a barrier between hydrogen and SiC.

(ii) It increases the diameter of the nanotubes, thus, decreasing their specific surface area.

(iii) Plugs the ends of the SiCNTs, thus preventing the hydrogen molecules from having access to the inner pores of the SiCNTs.

Keywords: hydrogen storage, SiC nanotubes, sorption, carbon nanotubes

---

References

[1] Dai H, Wong EW, Liu YZ, Fan SS, Lieber CM. Synthesis and characterization of carbide nanorods. Nature 1995;375:769-72.

[2] Zhang LD, Meng GW, Phillip F. Synthesis and characterization of nanowires and nanocables. Sci. Eng. A 2000;286:34-8.

[3] Yang TH, Chen CH, Chatterjee A, Li HY, Lo JT, Wu CT, Chen KH, Chen LC. Controlled growth of silicon carbide nanorods by rapid thermal process and their field emission properties. Chem. Phys. Lett. 2003;379:155-61.

[4] Honda S, Baek YG, Ikuno T, Kohara H, Katayama M, Oura K, Hirao T. SiC nanofibers grown by high power microwave plasma chemical vapor deposition. Appl. Surf. Sci. 2003;378:212-3.

[5] Yang ZX, Xia YD, Mokaya R. High surface area silicon carbide whiskers and nanotubes nanocast using mesoporous silica. Chem. Mater. 2004;16:3877

[6] Keller N, Pham-Huu C, Ehret G, Keller V, Ledoux MJ. Synthesis and characterization of medium surface area silicon carbide nanotubes. Carbon 2003;41:2131-9.

[7] Barghi SH, Tsotsis TT, Sahimi M. Chemisorption, physisorption and hysteresis during hydrogen storage in carbon nanotubes. Int. J. Hydrogen Energy 2014;39:1390-7.

Prediction of Methane Storage Performance of Metal-Organic Frameworks

Kutay Berk Sezginel[1,2], Alper Uzun[1,2] and Seda Keskin[1,2]
[1]*Department of Chemical and Biological Engineering, Koc University, Rumelifeneri Yolu, Sariyer, 34450, Istanbul, Turkey*
[2]*KÜTEM (Koc University TUPRAS Energy Center), Koc University Rumelifeneri Yolu, Sariyer, 34450, Istanbul, Turkey*

Natural gas, consisting mainly of CH_4, has the potential to replace petroleum and become the primary transportation fuel source. It has the highest H/C ratio among all hydrocarbons therefore emits less carbon compared to other fossil fuels.[1] It is also naturally abundant and represents more than half of the fossil fuel on earth. The major challenge of utilizing CH_4 as a fuel is the efficient room temperature storage and delivery due to its low volumetric energy density compared to gasoline. One of the promising methods to store high densities of CH_4 is adsorbed natural gas (ANG) since it offers a safer and relatively inexpensive solution compared to other methods such as compressed natural gas (CNG). In order to successfully implement ANG, adsorbent materials that can efficiently store and deliver CH_4 must be developed. For this purpose, various studies focused on porous materials such as zeolites, activated carbons, carbon nanotubes (CNTs) and metal-organic frameworks (MOFs). Among these porous materials MOFs have been examined extensively for their CH_4 storage applications. MOFs are a type of highly porous materials formed typically from metal clusters and organic linkers and they have shown high potential in a wide range of applications such as gas storage, chemical separations and catalysis.[2] Compared to other porous materials like zeolites, activated carbons, MOFs exhibit several advantages, such as ease of synthesis, stability, high porosity, and large surface area. Moreover, the fact that they are formed from a variety of modular building blocks provides the opportunity to obtain almost unlimited number of different materials by changing the combination of metal centers and organic linkers. Experimental identification of highly promising materials for CH_4 storage among a large number of MOFs is challenging and computational tools that can make accurate predictions about materials' uptake capacities are critical.

In this study, CH_4 storage in MOFs was investigated and relations between CH_4 storage, thermodynamic and structural properties of MOFs were studied in detail. CH_4 uptake and isosteric heat of adsorption (Q_{st}) values of 45 different MOF materials were calculated at ambient temperature and various pressures between 1 and 65 bar using Grand Canonical Monte Carlo (GCMC) simulations. The validity of these simulation results were tested by comparing the data with available experimental results for CH_4 uptake from the literature.[3] The simulations were performed at exactly same pressures with the experimental data and a very good agreement was observed with a regression coefficient of 0.972. The comparison of the simulation results and experimental data is given in Figure 1.

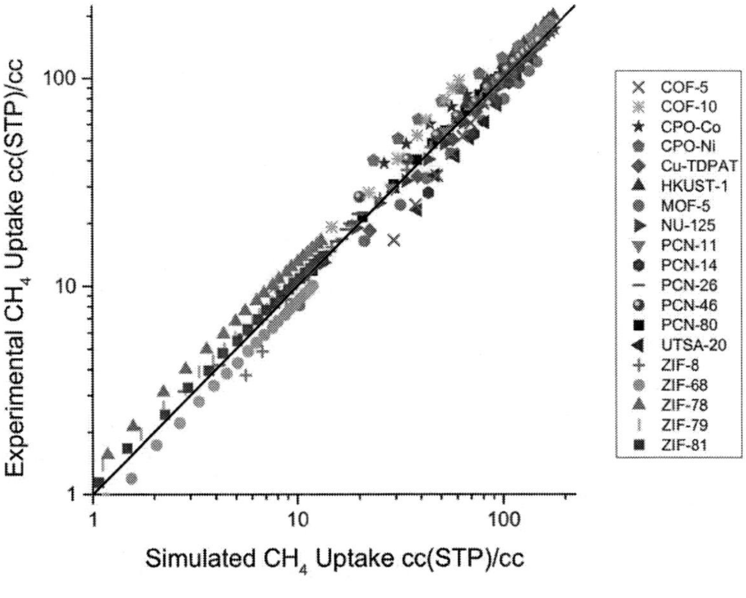

Figure 1. Comparison of experimental and simulated CH_4 uptake at 298 K in the range of 0.1 – 35 bar

In order to understand the most prominent factors affecting the CH_4 storage performance of MOFs, seven structural characteristics (crystallographic density, void, fraction, pore volume, surface area, pore limiting diameter, maximum pore diameter, and dominant pore diameter) and isosteric heat of adsorption (Q_{st}) values were calculated using computational tools. A quantitative structure-property relationship (QSPR) analysis was performed to determine which parameters are more correlated with CH_4 uptake at ambient temperature and pressures between 1 and 65 bar. Using the most correlated properties, multivariable linear models that predict CH_4 uptake were constructed making use of four variables. Models that predict CH_4 uptake at 5 and 35 bars were studied in detail considering the porous material storage target determined by the U.S. Department of Energy (U.S. DOE)[4] where 5 bar is considered as a relevant delivery pressure and 35 bar as a storage pressure at ambient temperature. It was found out that at delivery pressure of 5 bar crystal density, void fraction, pore volume and Q_{st} are the parameters that successfully represent the simulation result with a regression coefficient of 0.85 and at storage pressure of 35 bar void fraction, surface area, dominant pore diameter and Q_{st} are the parameters that successfully represents the simulation results with a regression coefficient of 0.92. The comparison of models with the simulation results for 5 and 35 bar are given in Figure 2.

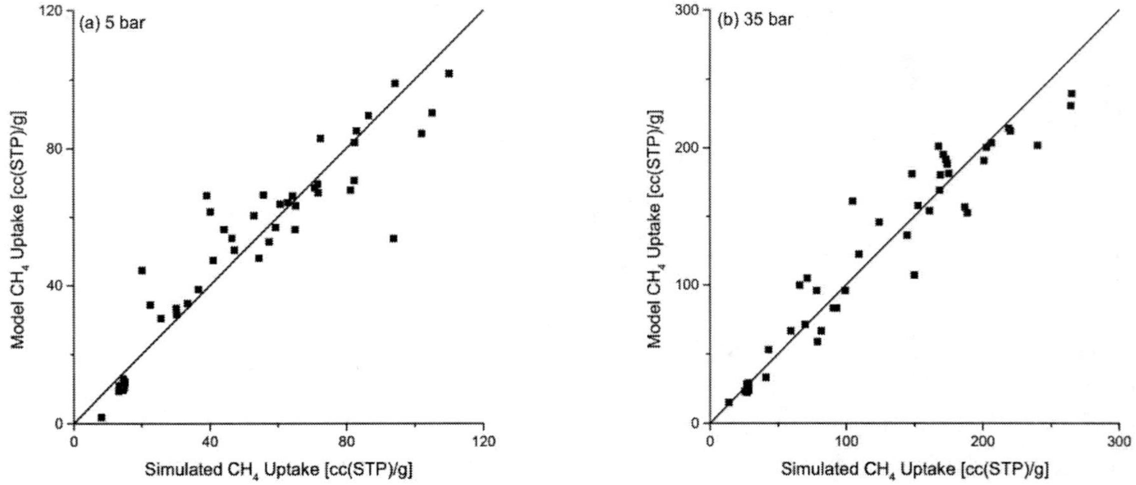

Figure 2. Comparison of models and simulations for CH_4 uptake at 298 K and (a) 5 bar (b) 35 bar

The model results were also compared with the experimental data and it was found that models can also represent experimental results very accurately. Moreover, the accuracy of the models were compared with the accuracy of previously proposed models by Fernandez et al. [5], He et al. [6], and Peng et al. [7] and it was shown that the models proposed in this study agrees better with the experiments since Q_{st} was included, which was not considered in previous modeling studies.

In conclusion, successful models were developed to predict CH_4 storage characteristics of MOFs using only structural characteristics and isosteric heat of adsorption values. These models can be helpful for the estimation of CH_4 uptake without compelling experimental adsorption measurements. Moreover, benefiting from these models valuable insights can be gained regarding which type of materials should the targets be for synthesis and application. However, it should be noted that for the final decision of which MOF to synthesize, other important factors such as cost of reagents and equipment, ease of experimental synthesis and time constraints should be carefully considered.

References

1. He, Y., et al., *Methane storage in metal-organic frameworks.* Chemical Society Reviews, 2014.
2. Furukawa, H., et al., *The Chemistry and Applications of Metal-Organic Frameworks.* Science, 2013. **341**(6149).
3. Mason, J.A., M. Veenstra, and J.R. Long, *Evaluating metal-organic frameworks for natural gas storage.* Chemical Science, 2014. **5**(1): p. 32-51.
4. Available from: https://arpa-e-foa.energy.gov/.
5. Fernandez, M., et al., *Large-Scale Quantitative Structure–Property Relationship (QSPR) Analysis of Methane Storage in Metal–Organic Frameworks.* The Journal of Physical Chemistry C, 2013. **117**(15): p. 7681-7689.
6. He, Y., et al., *A series of metal-organic frameworks with high methane uptake and an empirical equation for predicting methane storage capacity.* Energy & Environmental Science, 2013. **6**(9): p. 2735-2744.
7. Peng, Y., et al., *Methane Storage in Metal–Organic Frameworks: Current Records, Surprise Findings, and Challenges.* Journal of the American Chemical Society, 2013. **135**(32): p. 11887-11894.

389636 Excess Thermopower and the Theory of Thermopower Waves

Abstract:

Self-propagating exothermic chemical reactions can generate electrical pulses when guided along a conductive conduit such as a carbon nanotube. However, these thermopower waves are not described by an existing theory to explain the origin of power generation or why its magnitude exceeds the predictions of the Seebeck effect. In this work, we present a quantitative theory that describes the electrical dynamics of thermopower waves, showing that they produce an excess thermopower additive to the Seebeck prediction. Using synchronized, high-speed thermal, voltage, and wave velocity measurements, we link the additional power to the chemical potential gradient created by chemical reaction (up to 100 mV for picramide and sodium azide on carbon nanotubes). This theory accounts for the waves' unipolar voltage, their ability to propagate on good thermal conductors, and their high power, which is up to 120% larger than conventional thermopower from a fiber of all-semiconducting SWNTs. These results underscore the potential to exceed conventional figures of merit for thermoelectricity and allow us to bound the maximum power and efficiency attainable for such systems.

Novel Mesostructured Iron Oxides for Highly Reversible Lithium Storage

Jian Zhu; K.Y. Simon Ng; Da Deng

Department of Chemical Engineering and Materials Science, Wayne State University,

5050 Anthony Wayne Dr, Detroit, MI, United States, 48202

Carbon-alternative high-capacity negative electrode materials have to be developed to meet the increasing demand for mobile energy in personal electronic devices, electric vehicles and energy storage in grid. Transition metal oxides receive increasing attention as carbon-alternatives because of their much higher theoretical capacity and a totally different mechanism in lithium storage, as compared to graphite based on lithium intercalation with a moderate capacity of 372 mAh/g. In particular, Fe_2O_3 and Fe_3O_4 are very promising candidates with theoretical capacity of 1007 and 924 mAh/g, respectively. Iron oxides are outstanding in terms of low cost, abundance, ease of fabrication, and environmental benignity. However, the poor cyclability caused by volume change during the insertion /extraction of lithium ions and poor conductivity is the main obstacle. Our strategy is to design and tailor the structures of iron oxides at nanoscale and build mesostructures to improve their electrochemical performance. In this presentation, our latest results on the strategies to facilely prepare mesoporous iron oxides and the novel method to prepare additive-free electrodes will be reported. The dramatically improved electrochemical performance in reversible lithium storage will be discussed.

Mesoporous, Nanocrystalline SnO_2 Anodes for Excellent Lithium Ion Storage

Vinodkumar Etacheri, [1] *Gulaim A. Seisenbaeva,* [2,*] *Vadim G. Kessler* [2] *and* *Vilas G. Pol* [1,*]

[1]School of Chemical Engineering, Purdue University, 480 Stadium Mall Drive, West Lafayette, Indiana 47907-2100, United States, *vpol@purdue.edu

[2]Department of Chemistry, Swedish University of Agricultural Sciences, 75007, Uppsala, Sweden, *gulaim.seisenbaeva@slu.se

Keywords: Lithium ion batteries, Tin oxide, Mesoporous materials, Anode

INTRODUCTION

Lithium ions batteries are promising candidates for electrochemical energy storage due to their high energy density, enhanced rate capabilities and safety features.[1-3] They also find wide range of applications in portable electronic devices, implantable medical devices, electric vehicles etc. Use of graphite as anode (specific capacity of 372 mAh/g) is one of the main reasons responsible for the low specific capacity of Li-ion batteries, which is critical for many commercial applications including long range driving. Consequently, intensive research has been devoted to the development of high capacity anode materials. Due to superior theoretical capacity (781 mAh/g) than the conventional graphite anode, SnO_2 anode materials have attracted significant interest since the first report by Idota[4] et al. However, the critical factor that hinders the use of SnO_2 anodes is their poor cycling stability due to the huge volume change during the lithiation-delithiation (charge-discharge) process. A number of methods have been proposed to improve the cycling performance of SnO_2 anodes by minimizing the effects of drastic volume change. For instance, Li *et al.* reported that SnO_2 nanofibers effectively accommodated the volume change.[5] SnO_2 nanofibers anodes retained a good cycling stability at a high charge rate. Thin-films and nanosheets of SnO_2 were also studied for the same purpose.[6, 7] However, most of these electrode materials utilized complex and expensive synthesis methods. Introduction of solid alkoxide precursors into boiling water is a general very productive approach to prepare uniform mesoporous oxide materials, achieved through diffusion-controlled nucleation.[8] Here, we report synthesis and superior electrochemical performance of mesoporous SnO_2 nanoparticles through hydrolysis and heat treatment of tin tetraisopropoxide. This study also proved the requirement of higher order of mesoporosity for improving the cycling stability of nanocrystalline SnO_2 anodes with modified electrolytes.

MATERIALS AND METHODS

The preparation of ultrafine mesoporous SnO_2 was carried out by quick introduction of one gram of tin tetraisopropoxide (Multivalent Co UK) into 50 ml of boiling Milli-Q water with subsequent refluxing of the obtained milky dispersion for 30 min. The mixture was then left for precipitation overnight, the supernatant was removed by decantation and the solid residue was dried at 110° C in air for 3 hours, and calcined at 400 °C for 1 h; heating and cooling rates were 10 °C/min. X-ray diffraction (XRD) patterns of the samples (2θ = 10–70°) were collected using a Bruker D8 advance diffractometer equipped with a Lynx-Eye detector and parallel beam optics using Cu-Kα radiation (λ = 0.154184 nm). TGA investigations were made with Perkin-Elmer Pyris 1 instrument (heating rate 5 °C/min, temperature range 22°C – 600°C). TEM investigations were made using CM12 transmission electron microscope (Philips, Holland) with an accelerating voltage of 100 kV. SEM-EDS studies were carried out with a Hitachi TM-1000-μ-DeX tabletop scanning electron microscope. Textural characteristics (specific surface area, mean pore size, porous volume) were measured using nitrogen sorption at 77 K on a Quantachrome Autosorb 1 apparatus.

Electrodes for Li-ion batteries were prepared by mixing 80 % active material, 10 % carbon black and 10 % PVDF, and coated on a copper foil. These electrodes were assembled into coin-type using a Celgard 2500 polypropylene separator, and Li-foil as the counter electrode. The electrolyte consists of 1 M $LiPF_6$ in a 1:1 mixture of ethylene carbonate (EC)/dimethyl carbonate (DMC) and 3 volume % vinylene carbonate (VC). The electrochemical cells were assembled in a glovebox filled with high-purity argon (99.99%). Galvanostatic electrochemical measurements were performed using a Neware battery tester in the voltage range 10mV to 1.5 V.

RESULTS AND DISCUSSION

X-ray diffraction patterns (Fig. 1a) confirmed the phase purity of the as-prepared nanocrystalline cassiterite phase SnO_2. Particle size calculation using Scherrer equation and TEM images (Fig. 1b) confirmed the particle growth upon heat treatment. Calculated particle size of the as prepared and 400 °C heat-treated SnO_2 samples are 5 and 15 nm, respectively. N_2 adsorption-desorption isotherms of heat-treated SnO_2 sample demonstrated type IV with H1 type hysteresis,[9] and the as-prepared sample displayed no such hysteresis (Fig. 1c).

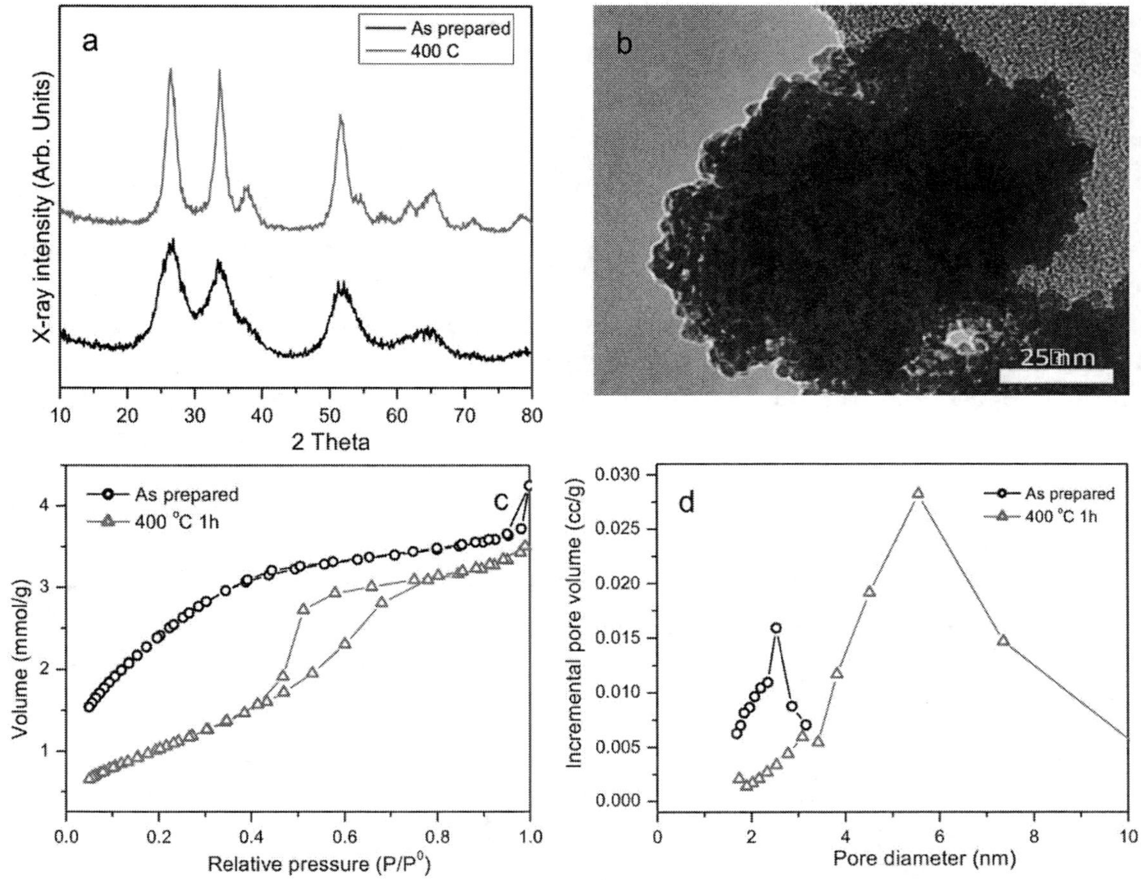

Fig. 1 (a) XRD patterns of as-prepared and heat-treated SnO_2 (400 °C in an air atmosphere); (b) TEM image of as-prepared SnO_2 sample; (c, d) N_2 adsorption-desorption isotherms and pore size distribution of as-prepared and heat-treated SnO_2 sample.

As-prepared SnO_2 sample has a high BET surface area of 204 m^2/g and average pore size of 2.2 nm. Whereas the surface area reduced to 87 m^2/g and pore size increased to 4.5 nm after 1 h heat-treatment at 400 °C (Fig. 1d). These results verified the particle growth during heat-treatment and generation of mesopores either in between the particles or on the individual SnO_2 nanoparticles.

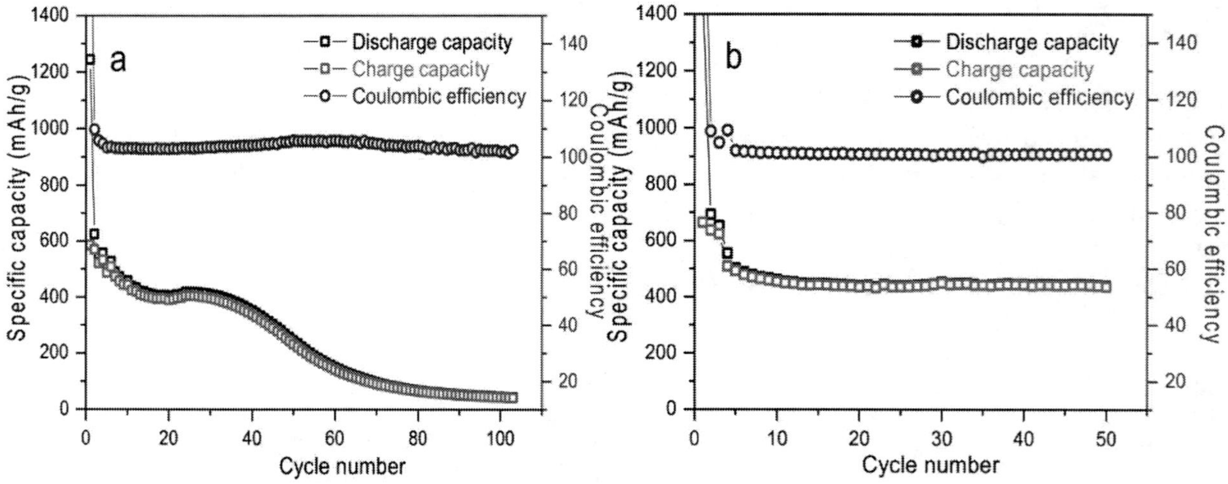

Fig. 2 Electrochemical performance of (a) as- prepared and (b) heat-treated SnO_2 nanoparticles

Despite of their small particle size and high surface area compared to the heat-treated SnO_2, as prepared samples displayed significant capacity fading upon prolonged cycling (Fig. 2a). On the other hand, low surface area SnO_2 heat-treated at 400 °C exhibited excellent capacity retention and coulombic efficiency on prolonged charge-discharge cycles (Fig. 2b). After 50 cycles at a charge-discharge rate of 2C (current density of 1560mA/g), mesoporous SnO_2 retained a specific capacity of 450 mAh/g, whereas the as-prepared showed a much lower capacity of 200 mAh/g. Electrochemical performance reported herein is superior to the previous reports on nanostructured SnO_2.[7,10] Vinylene carbonate used in this case also help to improve the cycling stability by forming polymeric surface films.[11] The excellent electrochemical performance of nanocrystalline SnO_2 heat-treated at 400 °C can be attributed to the high order mesoporosity, which facilitate superior electrode-electrolyte contact and strain relaxation during the lithiation-delithiation of Sn. Long term cycling performance and rate capabilities of the mesoporous nanocrystalline SnO_2 will be investigated.

CONCLUSIONS

In summary, mesoporous SnO_2 nanoparticles were successfully synthesized through the hydrolysis of tin tetraisopropoxide followed by heat treatment in air atmosphere. These nanoparticles demonstrated excellent Li-ion specific capacities and cycling stability. Electrolyte additive, vinylene carbonate was found to have a positive effect on the electrochemical cycling of current SnO_2 anode material. This study proved that high order of mesoporosity is highly beneficial for the electrochemical performance of inorganic SnO_2 nanoparticles.

REFERENCES

1. V. Etacheri, R. Marom, R. Elazari, G. Salitra and D. Aurbach, *Energy Environ. Sci.*, 2011, **4**, 3243-3262.
2. V. G. Pol and M. M. Thackeray, *Energy Environ. Sci.* , 2011, **4**, 1904–1912
3. V. Etacheri, J. E. Yourey and B. M. Bartlett, *ACS Nano*, 2014, 1491–1499.
4. Y. Idota, T. Kubota, A. Matsufuji, Y. Maekawa and T. Miyasaka, *Science*, 1997, **276**, 1395-1397.
5. N. Li, C. R. Martin and B. Scrosati, *Electrochem. Solid-State Lett.* , 2000, **3**, 316-318.
6. R. Retoux, T. Brousse and D. M. Schleich, *J. Electrochem. Soc.* , 1999, **146**, 2472-2476.
7. C. Wang, Y. Zhou, M. Ge, X. Xu, Z. Zhang and J. Z. Jiang, *J. Am. Chem. Soc*, 2010, **132**, 46–47.
8. G. A. Seisenbaeva, G. Daniel, J. M. Nedelec and V. G. Kessler, *Chem. Eur. J.*, 2014 DOI: 10.1002/chem.201402691.
9. V. Etacheri, G. Michlits, M. K. Seery, S. J. Hinder and S. C. Pillai, *ACS Appl. Mater. Interfaces*, 2013, **5**, 1663-1672.
10. S. H. Ng, D. I. d. Santos, S. Y. Chew, D. Wexler, J. Wang, S. X. Dou and H. K. Liu, *Electrochem. Commun.*, 2007, **9** 915–919.
11. D. Aurbach, K. Gamolsky, B. Markovsky, Y. Gofer, M. Schmidt and U. Heider, *Electrochim. Acta*, 2002, **47**, 1423-1439.

Electrochemical Study of Mesoporous Core-shell Lithium Titanate-Carbon Composite ($Li_4Ti_5O_{12}$-Carbon) with Controlled Microstructure as Anode in Lithium Ion Batteries for Wide Temperature Range

Chi-Ying Vanessa Li, Ching-Kit Ho,
Zhao-Feng Deng, Kwong-Yu Chan

Department of Chemistry, The University of Hong Kong, Pokfulam Road, Hong Kong

Integrate carbon and lithium titanate ($Li_4Ti_5O_{12}$) with appropriate structure control would enhance the rate performance of $Li_4Ti_5O_{12}$ as anode with improved electronic conductivity.[1-4] In this study, carbon with hollow core and mesoporous shells of controlled dimensions (in terms of diameter of core, thickness of the shell and diameter of the mesopores) were synthesized via silica hard-template method. $Li_4Ti_5O_{12}$-carbon composites were built through impregnation of $Li_4Ti_5O_{12}$ into the core-shell carbon. Cross-section TEM studies have been performed and proved that $Li_4Ti_5O_{12}$ particles are impregnated within the mesoporous carbon shell structure.

These $Li_4Ti_5O_{12}$-carbon composites demonstrate higher electronic conductivity, and enhanced electrochemical behavior for prolonged cycles at high discharge rate due to increased surface area of the active material, and improved ion transport pathways from possibly the length scale effect. The Coloubmic efficiency can achieve 98% and 96%, and superior long cycle life with high retention of 99% and 85% over 2000 cycles at 10C (2.2 A/g) and 20C (4.4 A/g), respectively. Outstanding electrochemical behavior is also observed for low temperature condition.

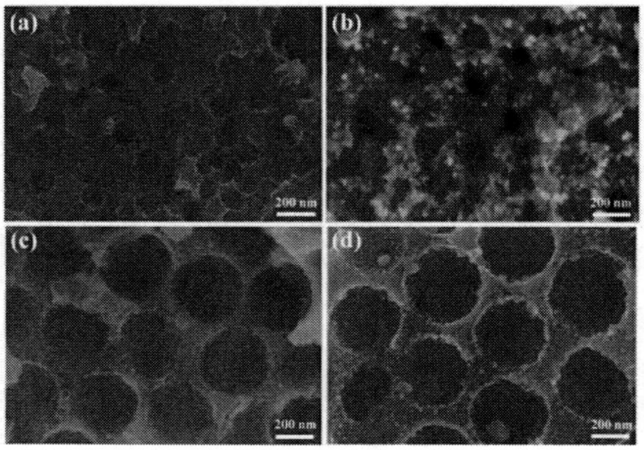

Fig. 1 SEM micrographs of HCMS carbon with shell thickness of ca. 50 nm and core diameter of (a) 150 nm (C-150) and (c) 350 nm (C-350); $Li_4Ti_5O_{12}$/carbon nanocomposite using (b) C-150 as substrate (LTO/C-150) and (d) C-350 as substrate (LTO/C-350).

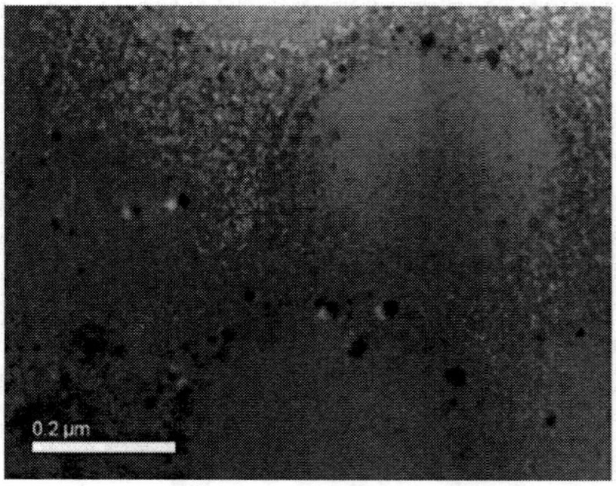

Fig. 2. Cross-sectional TEM image of $Li_4Ti_5O_{12}$/carbon nanocomposite.

Acknowledgement

This work was supported by Hong Kong Research Grants Council General Research Fund (HKU 700209P), Strategic Research Theme on Clean Energy, University Development Fund on Initiative on Clean Energy, and Innovation and Technology Fund of the Hong Kong SAR (ITS/076/11).

References

1. C.K. Ho, H. Yung, M. Zhou, C.Y.V. Li, K.Y. Chan, "Mesoporous Lithium Titanate-Carbon Composite with Controlled Microstructure", Abstract #1148, 224th ECS Meeting, October 2013, San Francisco, U.S.A.
2. S.W. Ting, C.Y.V. Li, H. Yung. C.K. Ho and K.Y. Chan, "Structural and Electrochemical Studies of Mesoporous $Li_4Ti_5O_{12}$-TiO_2 Composite Spheres as Anode Material for Lithium ion Batteries, 222nd ECS Meeting, October 2012, Honolulu, Hawaii, U.S.A.
3. C.Y.V. Li, S.W. Ting, Z. Yang, W. Zhuang, X. Lu, H. Yung and K.Y. Chan, "Lithium Titanate Prepared from Mesoporous TiO_2 Fiber as Anode Material for Lithium Ion Batteries, , 222nd ECS Meeting, October2012, Honolulu, Hawaii, U.S.A.
4. X. Zheng, C. Dong, B. Huang, M. Lu, Int. J. Electrochem. Sci., 7 (2012) 9869 – 9880.

Design and Fabrication of High-Performance Electrodes for Li-ion Batteries Based on Sequential Multilayer Electroplating

Our study focuses on developing rationally designed and deterministically engineered structures to be used as electrodes in batteries with rapid charging capabilities. These high-power electrodes are achieved by the minimization of certain inherent resistances taking place during the transportation process of ionic and electronic species within the battery. For the realization of these electrodes, electrochemical as well as microfabrication techniques have been utilized. Diffusion-limited models have been adopted to concurrently optimize the liquid- and solid-state ion transport for high power capabilities by determining the optimum active material thickness as well as the inter-layer gap distance within the multilayer electrode. Robotically-assisted electroplating setup has been utilized to fabricate the modeled multilayer structures with alternating layers through a standard photoresist mold enabled by photolithography techniques. By adjusting the current density and electrodeposition time, structures composed of sub-micron-thick individual layers are realized whose total thicknesses reach up to several hundreds of microns. 100 pairs nickel (Ni) and copper (Cu) layers are electroplated from their respective plating baths in an alternating fashion under galvanostatic conditions. Following the fabrication of multilayer Ni/Cu structures of desired thickness, the first mold is stripped and a second photoresist molding process is carried out to electroplate Ni anchors on the specific regions of the multilayer structure. These Ni anchors serve as both supporting columns for the suspended structure and electrical connection between the Ni layers following the complete and selective removal of sacrificial Cu layers. The resulting Ni backbone structure exhibits a large electrode surface area, as well as serves a highly conductive current collector. To demonstrate the contribution of these architectures to the overall performance of the electrode, a conformal film of nickel-tin (NiSn) serving as the active material is electrodeposited onto the Ni backbone in the last step of the fabrication process. To ensure a conformal coating, pulse plating techniques are utilized. This conformal coating of the active material with sub-micron thickness provides a reduced diffusion and conduction path lengths for the ionic and electronic species, respectively, and hence, leads to high power capabilities. To confirm that, galvanostatic charge and discharge tests are conducted at rates up to 60 C (i.e., charging completely in 1 minute). The electrodes exhibit remarkable areal capacities of 2 mAh cm^{-2} and are able to deliver more than 50% of their 1 C capacities at rates as high as 60 C.

386988 Electrospun Mesoporous Nanofiber As Battery Application

Abstract:

We fabricated a polymer/ceramic hybrid nanofiber by electrospinning poly acrylonitrile(PAN)/polysilsesquioxane (up to 70/30 weight % ratio) composite solution. By tuning the calcination condition, either well-ordered mesoporous silica or carbon nanofibers are produced from this electrospun hybrid nanofiber. Visual characterization using transmission electron microscopy(TEM) and the pore size distribution measured using the BET surface analyzer confirmed the presence of the mesoporous structures made. These mesoporous materials can provide extremely high pore density and facile ionic transportation, which are useful traits as battery electrodes. The mesoporous silica nanofiber, after reduction, can turn into mesoporous silicon nanofiber which has great potential in Li-Ion battery anode. The mesoporous carbon nanofiber, on the other hand, shows good promise as Li-Sulfur battery material. These materials exhibit higher performance as battery materials after highly-conductive carbon nanotubes (CNT) or graphene nanoribbons (GNR) are incoporated.

Facile Synthesis of Strongly Coupled Carbon Nanofiber-Metal Oxide Coaxial Nanocables as High Performance Anode Materials for Lithium-ion Batteries

Genqiang Zhang, MPA-CINT, Los Alamos National Lab., Los Alamos, NM, 87544; Email:gqzhang@lanl.gov

Abstract

Lithium-ion batteries (LIBs) have been considered as one of the most promising electric energy storage systems due to the various merits of high voltage, high capacity, low cost and environmental friendliness. However, there are still several challenging issues in LIBs to be overcome in order to fulfil the requirements as high-performance power sources in future market, including higher capacity, lower cost, longer cycle life and better rate performance. Nanostructure engineering has been demonstrated as a powerful and effective strategy in the rational design of new electrode materials with optimized performance through both morphological and compositional solutions. For example, various nanostructures with controlled morphologies including hollow spheres/cubes, nanowires/rods as well as nanoplates could achieve significantly enhanced lithium storage properties due to the size and shape effects. However, the intrinsic properties including low electrical conductivity and poor mechanical stability of most oxide electrode materials, leading to unsatisfactory cycling stability and rate performance, still seriously limit their applications. Recently, an emerging concept of hybrid nanostructures has attracted tremendous attention since better performance could be expected in such architectures. Coaxial nanocables, as one of the most interesting hybrid nanostructures, hold the great potential for simultaneously resolving the problems of poor electrical conductivity and weak mechanical stability with rational design and careful choice of core and shell components. However, there are not too many such reports showing the attractive properties of nanocable structures for lithium storage, probably due to the lack of appropriate synthesis methods.

In this work, we have successfully synthesized CNF@MnO and CNF@$CoMn_2O_4$ coaxial nanocables with considerably enhanced lithium storage performance through a facile two-step strategy. The method involves the polyol process for the synthesis of metal-glycolate layer on the surface of the CNFs and the subsequent thermal ennealing treatment undr N_2 protection. These two nanocable electrodes exhibit remarkable lithium storage properties in terms of high specific capacity, long cycle life and superior rate performance. For example, at a current density of 1000 mA/g, the CNF@$CoMn_2O_4$ nanocable electrode can deliver a high capacity of about 655 mAh/g and can last for at least 300 cycles, which is remarkable. The enhanced electrochemical performance could be attributed to several advantages of the smart coaxial nanocable configuration, which can effectively alleviate the volume change, prohibit the nanoparticle aggregation and facilitate the electron transport. Such high-performance nanosctructures with large-scale production might hold great potential for the fabrication of high energy and power density lithium-ion batteries.

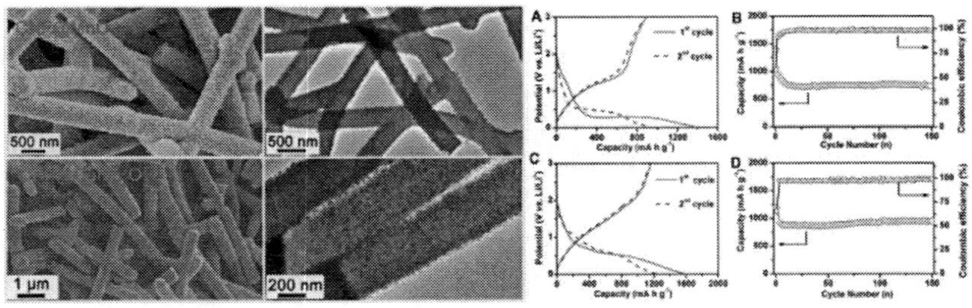

Typical morphologies and lithium storage properties as anode for LIBs

References:

1. G. Q. Zhang, H. B. Wu, H. E. Hoster, X. W. Lou,* ***Energy & Environmental Science***, *7*, 302 – 305 (**2014**).
2. L. Zhang, G. Q. Zhang, H. B. Wu, L. Yu, X. W. Lou,* ***Advanced Materials***, *25*, 2589 – 2593 (**2013**).

375082 New Approaches for the Processing of Oxygen Reduction Fuel Cell Catalysts

Abstract:

Hydrogen fuel cell technology has come a long way. Now this technology reaches a level that many technical hurdles have been overcome for vehicle applications. One major remaining challenge is the catalysts that are required for reducing oxygen at cathode. Various Pt and its alloy nanostructures have been examined for their use as highly active and stable oxygen reduction reaction (ORR) catalysts. In this talk, I will focus on the latest understanding of nucleation and growth of Pt nanoparticles in non-aqueous solutions, aiming at the design of advanced ORR electrocatalysts with high activity and selectivity. I will focus on the following topical areas that we have worked on: Preparation of nanoparticles, especially gas reducing agent in liquid solution (GRAILS) method for the preparation of shape- and composition-controlled Pt multimetallic nanostructures; ligand chemistry in the design and controlled synthesis of Pt and its alloy nanostructured catalysts; new potentials for using *in situ* liquid flow cell transmission electron microscope to study the formation of complex metallic nanoparticles, and applications of these nanoparticles as ORR electrocatalysts.

367449 Novel Pt-Based Catalysts for the Oxygen Reduction Reaction

Abstract:

Platinum-based catalysts are widely used in many industrial applications, including the proton-exchange membrane fuel cells. In addition to size, the shape and composition of a Pt-based nanocrystal also provide very sensitive knobs for tailoring its catalytic and electrocatalytic properties. A number of chemical routes have been developed to produce Pt-based nanocrystals with well-controlled sizes, shapes and elemental compositions to improve their performance in catalyzing various reactions, including the oxygen reduction reaction (ORR). In this talk, I will discuss a number of new developments from my own research group with regard to the design and synthesis of Pt-based catalysts with remarkable activity and/or durability for ORR. The strategies include the deposition of Pt ultrathin shells (a few atomic layers thick) on facet-controlled Pd nanocrystals, as well as the formation of Pt-Ni alloy octahedra with controllable sizes and compositions.

370182 Oxidation of Platinum Nickel Nanowires to Limit Displacement and Improve the Durability Characteristics of Oxygen-Reducing Electrocatalysts

Abstract:

Oxidation of Platinum Nickel Nanowires to Limit Displacement and Improve the Durability Characteristics of Oxygen-Reducing Electrocatalysts

Shaun M. Alia,[1] Svitlana Pylypenko,[2] Arrelaine Dameron,[1] K. C. Neyerlin,[1] Shyam S. Kocha,[1] and Bryan S. Pivovar[1,]*

[1] National Renewable Energy Laboratory, [2]Colorado School of Mines

National Renewable Energy Laboratory

Chemical and Materials Science Center

Golden, CO 80401

bryan.pivovar@nrel.gov

The cost of platinum (Pt) catalysts remains a significant barrier in the commercial development of proton exchange membrane fuel cells (PEMFCs). Carbon supported Pt nanoparticles (Pt/HSC) are typically used in PEMFCs due to their high surface area and moderate oxygen reduction reaction (ORR) mass activity. Extended Pt structures have previously been found to have specific activity and durability benefits to Pt nanoparticles.[1-3] Although extended catalysts can produce high specific activities, they generally lack the surface area to produce high mass activity. Catalysts formed by galvanic displacement are poised to produce high mass activities for ORR by maintaining the high specific activities demonstrated by extended Pt surfaces and by allowing for the deposition of thin Pt layers. Nickel (Ni) nanowires were previously used as a template for galvanic displacement, producing Pt coated Ni nanowires with surface areas as high as 91.3 m^2 g_{PGM} ^{1} and mass ORR activities as high as 917 mA mg_{PGM} ^{1}, 3.1 times greater than Pt/HSC (in rotating disk electrode, RDE, half-cell measurements).

Ni nanowires and Pt coated Ni nanowires were treated in oxygen at elevated temperatures. By heat treating in oxygen: the effects of an increasing Ni nanowire oxide layer on Pt displacement were examined; and durability losses (surface area and activity) of Pt coated Ni nanowires were reduced. As-received Ni nanowires contain a thin oxide layer and a Ni metal core. Heat treatment of the Ni nanowires with oxygen served to grow the oxide layer, increasing the oxide content near the nanowire surface, and oxidizing the metal nanowire core (to NiO) at higher temperatures. Increasing the oxide layer served to limit and eventually shut off Pt displacement.

Although untreated Pt coated Ni nanowires produce high mass activities, losses (similar to Pt/HSC) were observed following RDE durability testing (30,000 potential cycles, 0.6 1.0 V vs. RHE, similar protocol to DOE working group).[5,6] While extended Pt structures have previously been found to offer durability benefits to Pt/HSC, the benefit is likely lessened by the presence of Ni metal. By heat treating Pt coated Ni nanowires in oxygen, the oxide layer thickness was increased thereby stabilizing the catalysts in potential cycling. Treated catalyst produced initial surface areas and activities that were lower than the untreated Pt coated Ni nanowires, presumably since Ni near the surface could not be removed electrochemically during break-in. With potential cycling, however, treatment of the nanowires dramatically improved retention of surface area and activity. Following durability testing, treated Pt coated Ni nanowires produced a mass ORR activity of 680 mA mg_{PGM} ^{1}, or a 9 % improvement of initial activity. Compared to conventional Pt electrocatalysts and fully displaced Pt nanotubes, Pt coated Ni nanowires appear to provide significant activity and durability advantages.

References

1. Debe, M.; Department of Energy, U. S., Ed. http://www.hydrogen.energy.gov/pdfs/review08/fc_1_debe.pdf, 2008.

2. Chen, Z.; Waje, M.; Li, W.; Yan, Y. Angew. Chem. Int. Ed. 2007, 46, 4060-4063.

3. Papandrew, A. B.; Atkinson, R. W.; Goenaga, G. A.; Wilson, D. L.; Kocha, S. S.; Neyerlin, K. C.; Zack, J. W.; Pivovar, B. S.; Zawodzinski, T. A. ECS Transactions 2013, 50, 1397-1403.H. A. Gasteiger, S. S. Kocha, B. Sompalli, F. T. Wagner, *Applied Catalysis B-Environmental* **2005**, *56*, 9.

4. Alia, S. M.; Larsen, B. A.; Pylypenko, S.; Cullen, D. A.; Diercks, D. R.; Neyerlin, K. C.; Kocha, S. S.; Pivovar, B. S. ACS Catalysis 2014, 4, 1114-1119.

5. M. Uchimura, S. Sugawara, Y. Suzuki, J. Zhang, S. S. Kocha, *ECS Transactions* **2008**, 16, 225.

6. S. S. Kocha, *Electrochemical Degradation: Electrocatalyst and Support Durability*. In M. Mench, E. C. Kumbur, T. N. Veziroglu, Polymer Electrolyte Fuel Cell Degradation, pp. 89-185, 2011.

1

Green Microwave-Assisted Synthesis of AuPd/C and AgPd/C Catalysts for Methanol Electrooxidation in Alkaline Media

UNIVERSIDAD AUTÓNOMA DE CHIHUAHUA

Facultad de Ciencias Químicas

cimav

PRESENTER:

B. Ch. E. Ana Isabel Casas Hidalgo (UACH)

Advisors:

Ph. D. Hugo Ramos Sánchez (UACH)

M. S. Manuel Román Aguirre (CIMAV)

Colaborator:

Ph. D. Edgar Valenzuela Mondaca (UABC)

2

Microwave-Assisted synthesis of metallic Nanoparticles (NP)

- Microwave-assisted synthesis using natural extracts has gained interest as an environmentally friendly, time and cost effective methodology to produce metallic NP that may be used as the active phase of catalysts.[1-3]

1. Baruwati B., Varma R. S., *High Value Products from Waste: Grape Pomace Extract- A Three-in-one Package for the Synthesis of Metal Nanoparticles*. ChemSusChem (2009), p. 1041-1044.
2. Mittal A. K., Chisti Y., Banarjee U. C., *Synthesis of metallic nanoparticles using plant extracts*. Biotechnology Advances (2013), p. 346-356.
3. Nadagouda M. N., Varma R. S., *Green synthesis of silver and palladium nanoparticles at room temperature using coffee and tea extract*. Green Chemistry (2008), p. 859-862.

Synergetic effect of Au and Ag in catalytic activity

3

- It is well known that Pd alone shows good activity for the methanol oxidation reaction (MOR) in alkaline media. (4, 5)
- In recent years many studies have reported that the performance of catalysts for MOR in alkaline media can be significantly enhanced by the addition of a second metal to the active phase to produce a Pd-based alloy. (6, 7)

4. Singh, R.N., A. Singh, and Anindita, *Electrocatalytic activity of binary and ternary composite films of Pd, MWCNT and Ni, Part II: Methanol electrooxidation in 1 M KOH.* International Journal of Hydrogen Energy (2009), p. 2052-2057.
5. Wang, X., W. Wang, Z. Qi, C. Zhao, H. Ji, y Z. Zhang, *High catalytic activity of ultrafine nanoporous palladium for electro-oxidation of methanol, ethanol, and formic acid.* Electrochemistry Communications (2009), p. 1896-1899.
6. Liu, Z., X. Zhang, y L. Hong, *Physical and electrochemical characterizations of nanostructured Pd/C and PdNi/C catalysts for methanol oxidation.* Electrochemistry Communications (2009), p. 925-928.
7. Wang, X., B. Tang, X. Huang, Y. Ma, y Z. Zhang, *High activity of novel nanoporous Pd–Au catalyst for methanol electro-oxidation in alkaline media.* Journal of Alloys and Compounds (2013), p. 120-126.

Justification

4

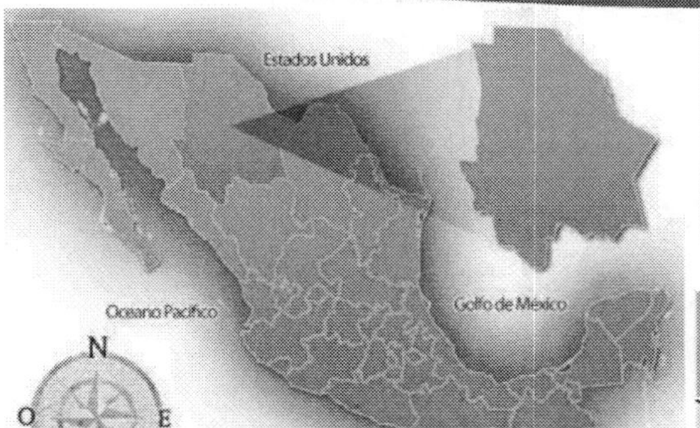

Table 1. Global position of Mexico in terms of production of gold and silver over the past 2 years(6)

Year / Metal	2012	2013
Silver	1°	1°
Gold	10°	11ª

Table 2. Chihuahua state production of pecan nut over the past 2 years(7)

Year / Product	2012	2013
Pecan nut* (Ton)	54, 000	64, 000
% Nacional Production	62%	58.1%

**Weight of the hole product*

6. Cámara Minera de México (CAMIMEX).
7. Comité Mexicano del Sistema Producto Nuez (COMENUEZ)

What we propose...

- Synthesize bimetallic AuPd/C and AgPd/C electrocatalysts using aqueous nutshell (*Carya illinoinensis*) extract and a microwave-assisted methodology.

Experimental Phase

1. Aqueous nutshell extract obtention

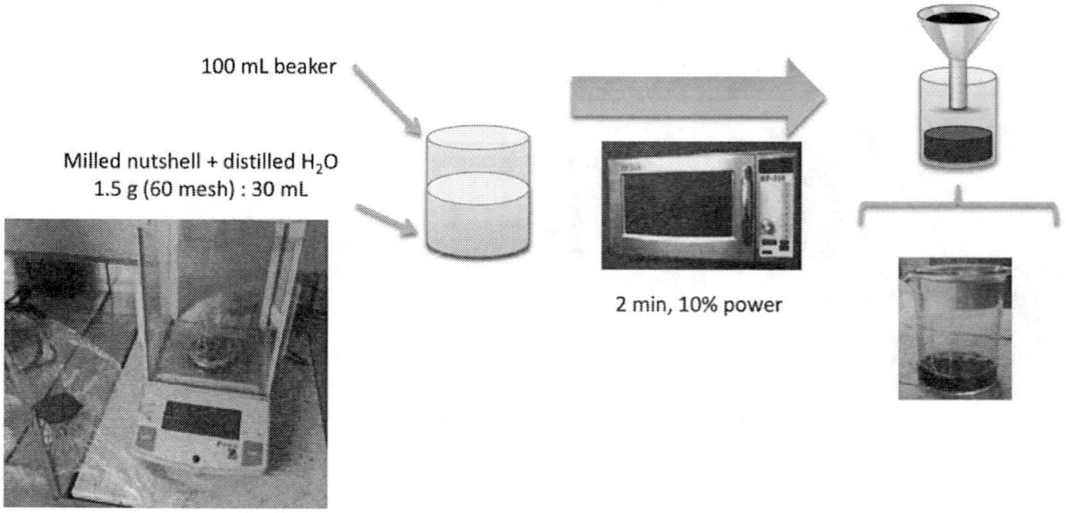

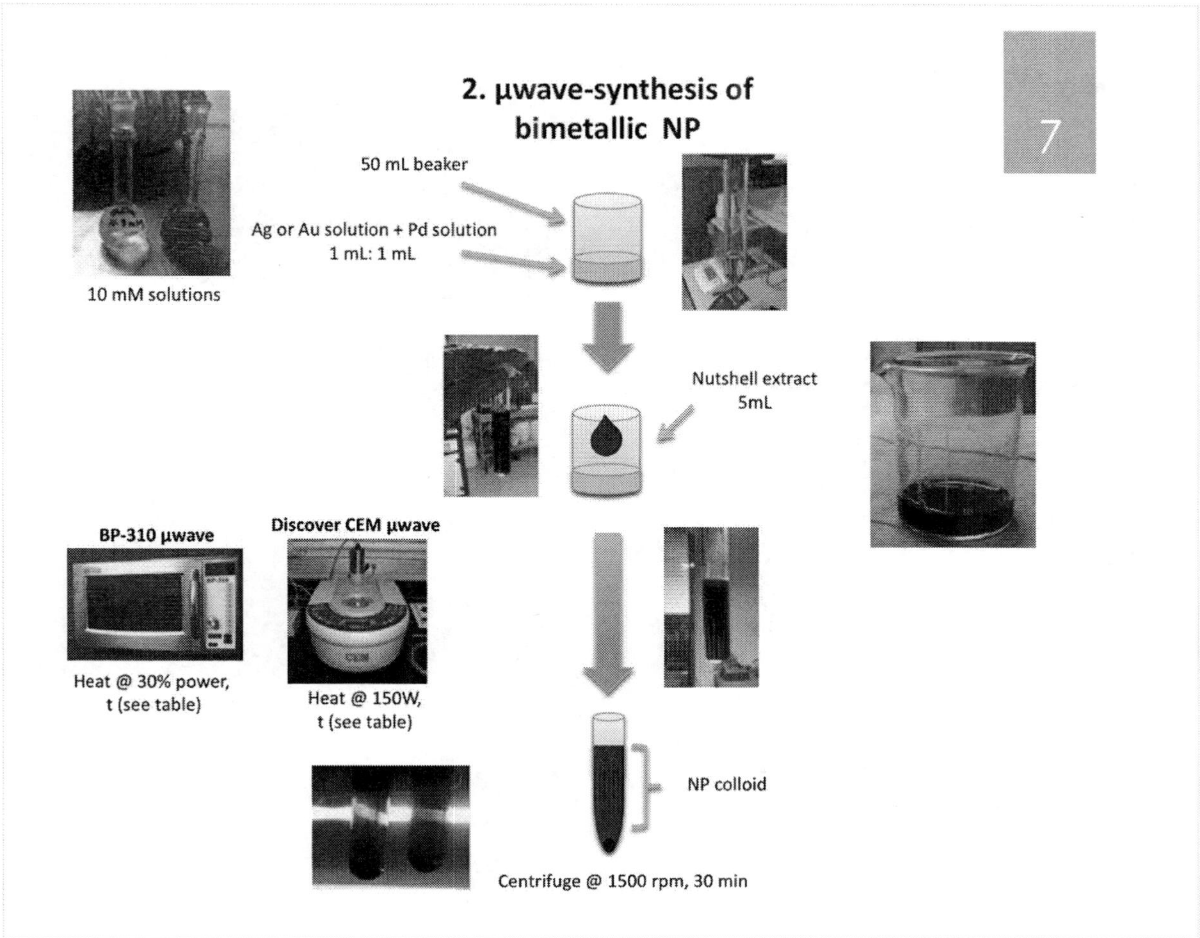

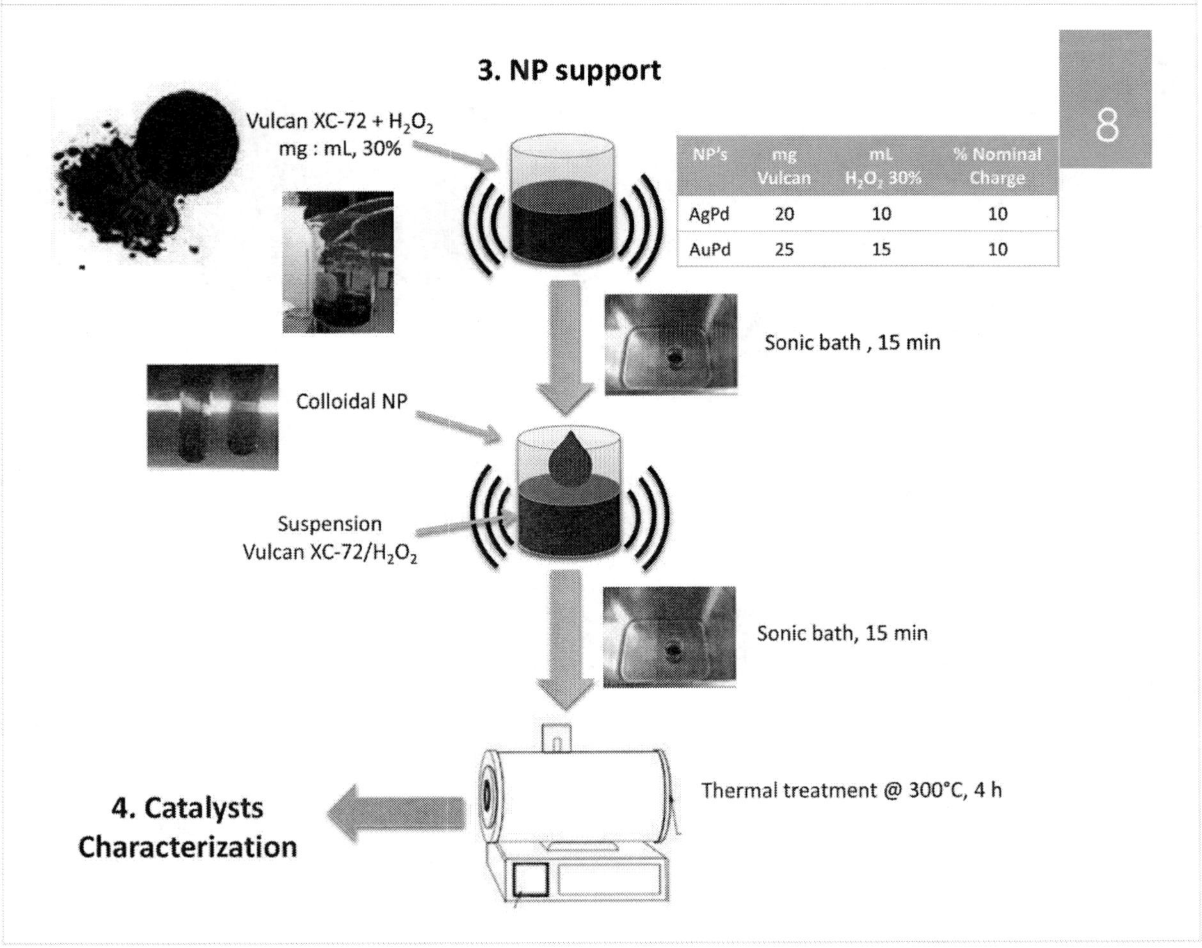

NP's	mg Vulcan	mL H_2O_2 30%	% Nominal Charge
AgPd	20	10	10
AuPd	25	15	10

Results

9

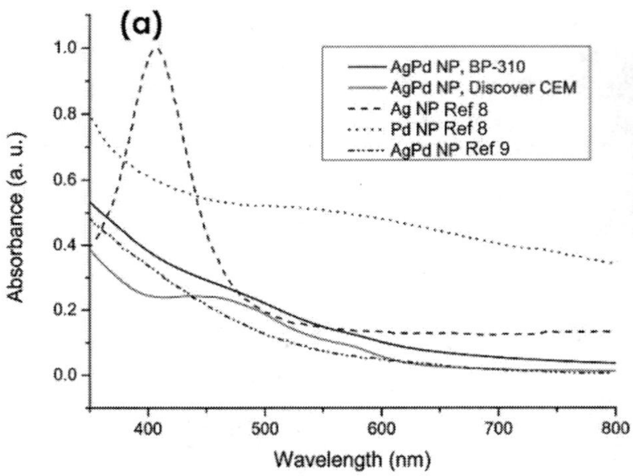

Figure 1. UV-Vis spectra of synthesized bimetallic (a) AgPd NP and (b) AuPd NP compared with spectra from literature

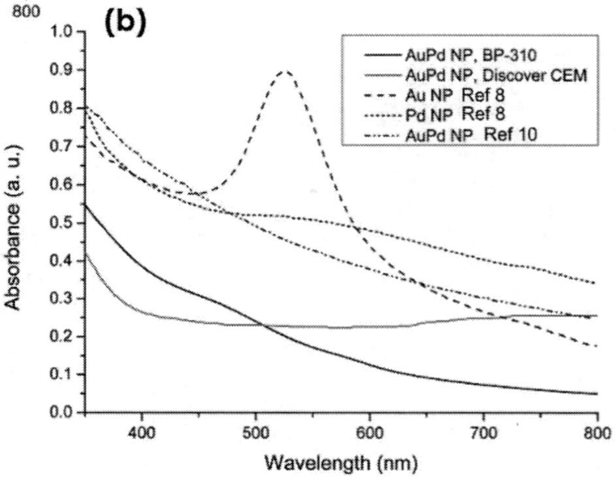

8. Panigrahi, S., *et al.*, *Sugar assisted evolution of mono- and bimetallic nanoparticles.* Colloids and Surfaces A: Physicochemical and Engineering Aspects, (2005), p- 133-138.
9. Yang, C.-C., C.-C. Wan, and Y.-Y. Wang, *Synthesis of Ag/Pd nanoparticles via reactive micelles as templates and its application to electroless copper deposition.* Journal of Colloid and Interface Science, (2004), p. 433-439.
10. Hou, W., N.A. Dehm, and R.W.J. Scott, *Alcohol oxidations in aqueous solutions using Au, Pd, and bimetallic AuPd nanoparticle catalysts.* Journal of Catalysis, (2008), p. 22-27.

Results

10

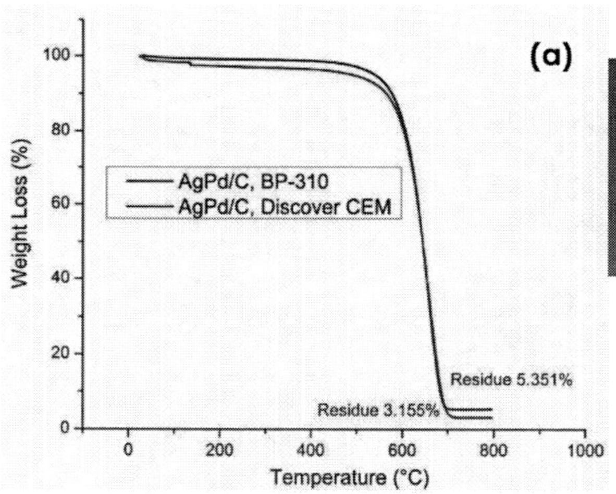

Figure 2. TGA analysis of (a) AgPd/C and (b) AuPd/C synthesized catalysts.

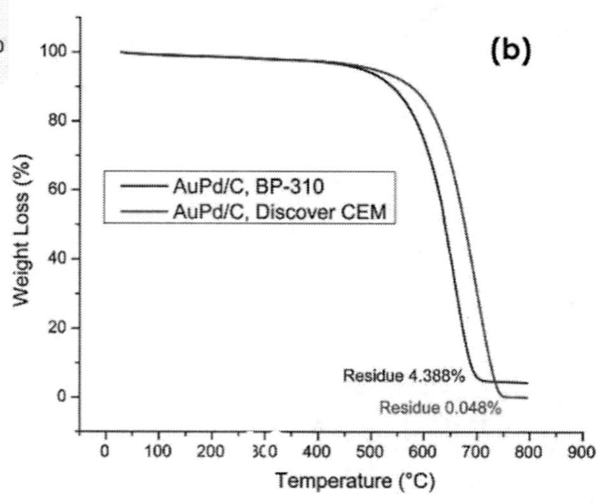

Results

11

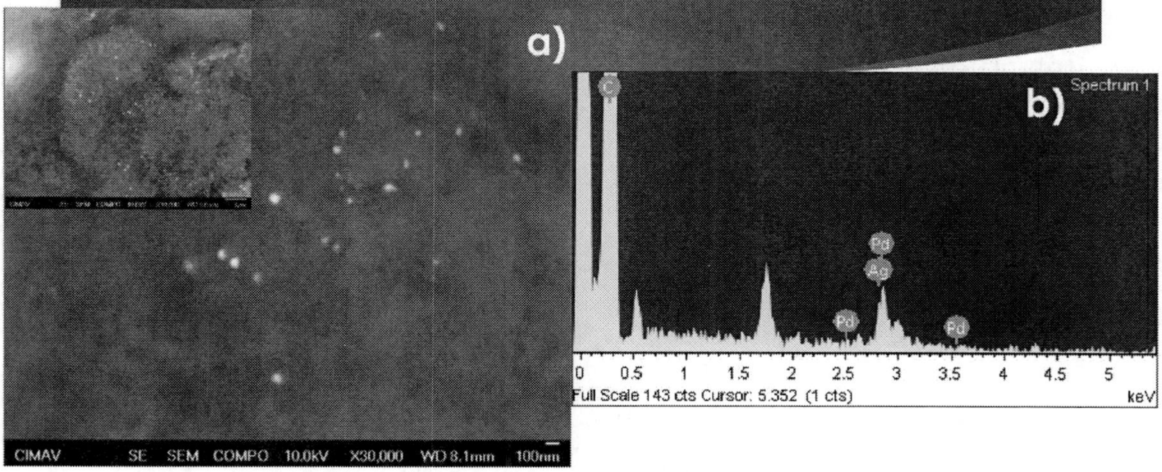

Figure 3. SEM micrograph (a) and EDS analysis (b) for the AgPd/C synthesized in Discover CEM μwave. Insert in graph(a) displays the image at 1 μm scale.

Results

12

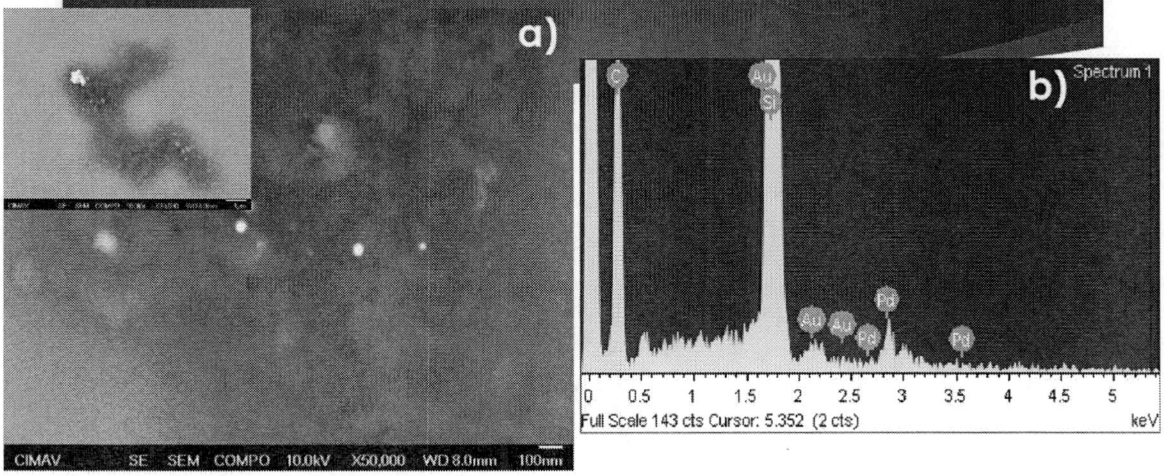

Figure 4. SEM micrograph (a) and EDS analysis (b) for the AuPd/C catalyst synthesized in BP-310 μwave. Insert in graph(a) displays the image at 1 μm scale.

Results

13

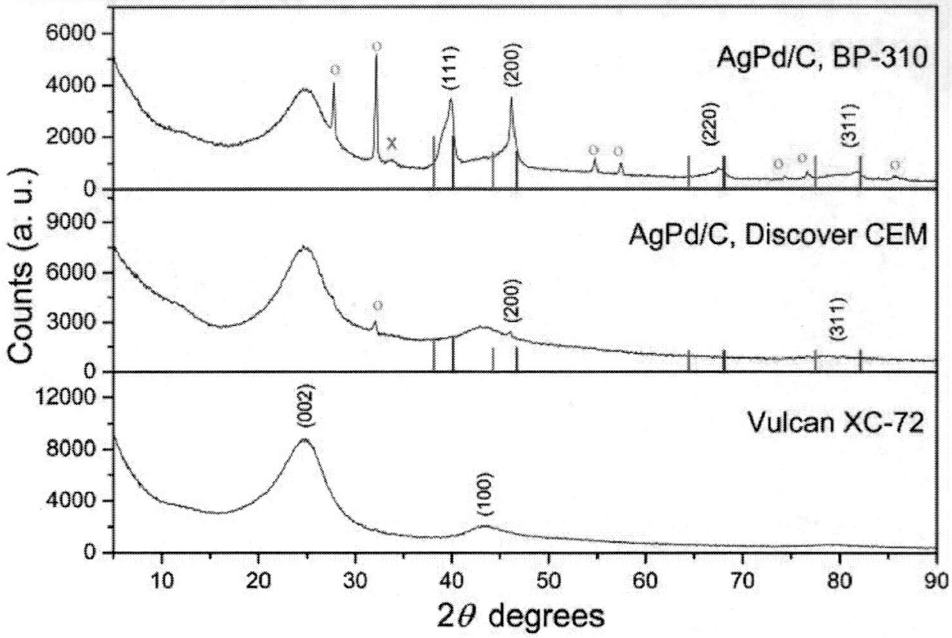

Figure 5. XRD patterns of AgPd/C catalyst. Ag reference (JCPDS 04–0783, blue line), Pd reference (JCPDS 05–0681l, gray line); (o) peaks related to PdO, (x) peaks related to Ag_2O.

Results

14

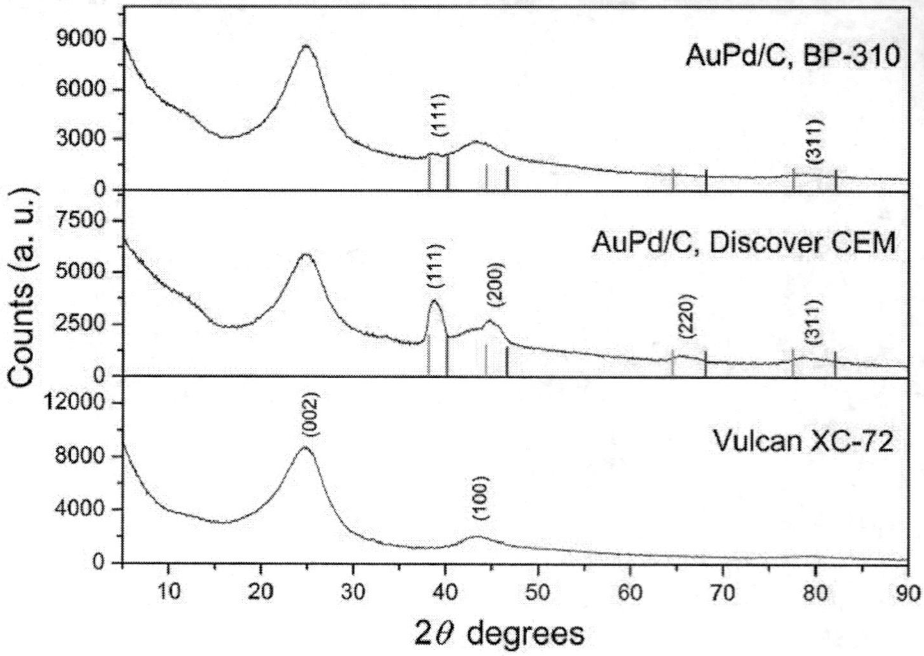

Figure 6. XRD patterns of AuPd/C catalyst. Au reference (JCPDS 04–0783, orange line), Pd reference (JCPDS 05–0681l, gray line.

Results

Table 3. Particle size and alloying degree of the synthesized catalysts

Microwave used for synthesis	Catalyst	Particle size* (nm)	Alloying degree†	Molar Fraction (XRF)
BP – 310	AgPd/C	7	$Ag_{24}Pd_{76}$	$Ag_{52}Pd_{48}$
	AuPd/C ♦	5	$Au_{72}Pd_{28}$	$Au_{46}Pd_{54}$
Discover CEM	AgPd/C ♦	21	$Ag_{27}Pd_{73}$	$Ag_{26}Pd_{74}$
	AuPd/C	5	$Au_{82}Pd_{18}$	$Au_{78}Pd_{22}$

* Calculated using Scherrer's equation and XRD data

† Calculated using Vegard's law XRD data

♦ Samples electrochemically characterized

$$*d = \frac{k\lambda}{\beta cos\theta}$$

$$\dagger\ a_{alloy} = a_{Pd} + kx_M$$

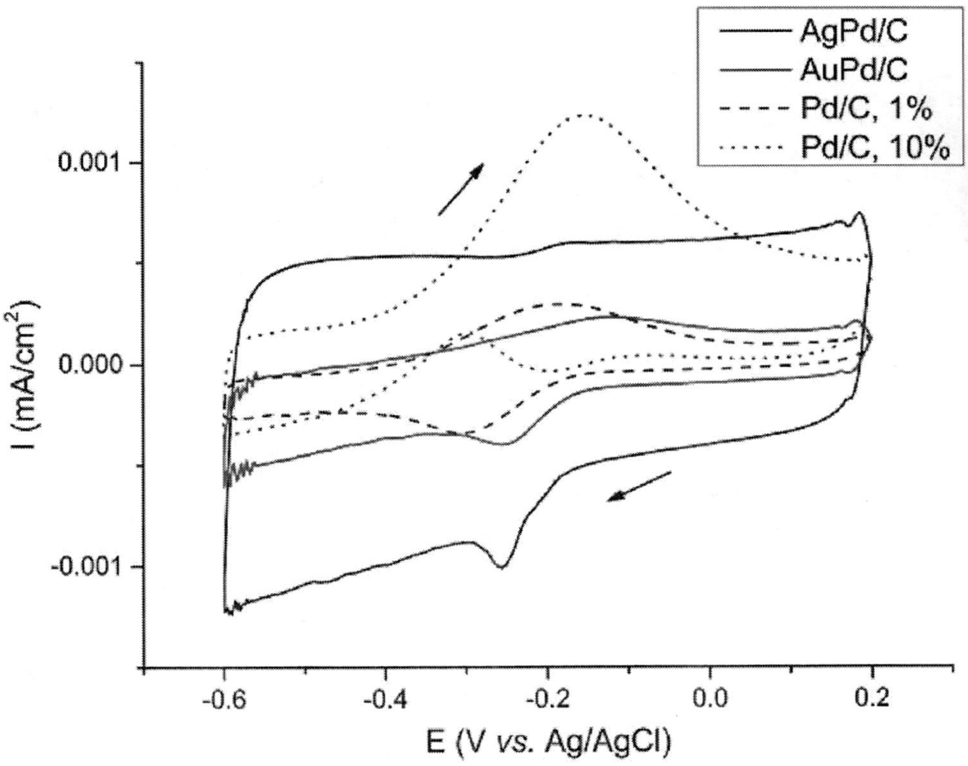

Figure 7. Cyclic Voltammetry for the AuPd/C and AgPd/C synthesized catalysts and commercial 1% Pd/C and 10% Pd/C catalysts in 1.0 M CH_3OH/0.5 M NaOH solution at a scan rate of 10 mV/s.

17

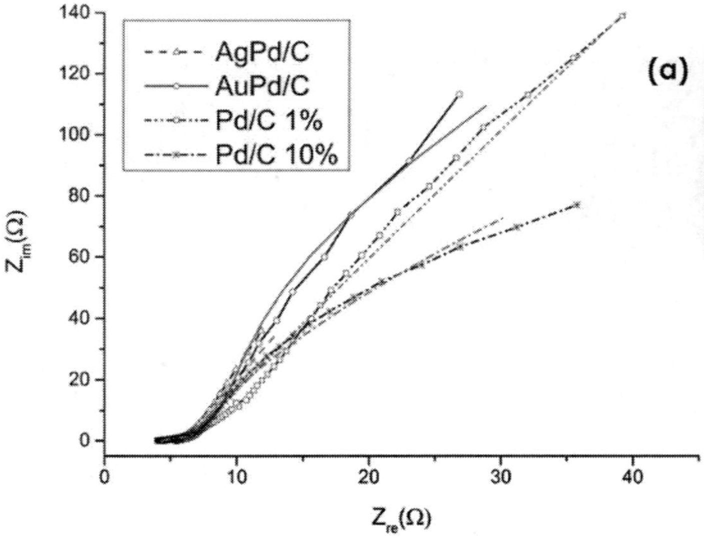

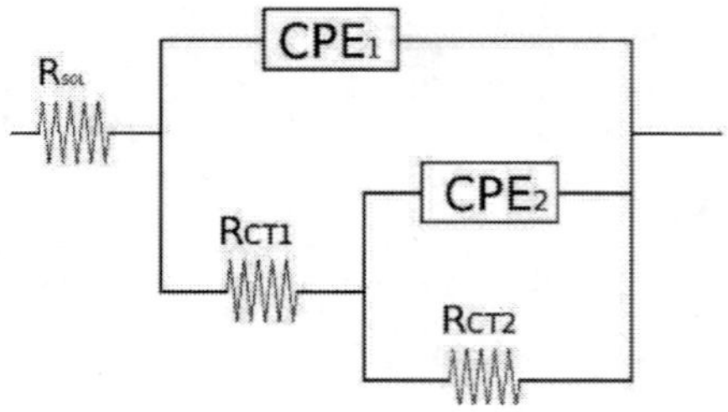

Table 4. Charge Transfer values (R_{CT}) for the analyzed catalysts

Catalyst	Charge Transfer (R_{CT}) Ω
AgPd/C	6.8 E11
AuPd/C	44742
Pd/C al 1%	5.1 E15
Pd/C al 10%	865

Figure 8. EIS analysis(a) and equivalent circuit (b) simulated for the analyzed AuPd/C, AgPd/C and commercial 1% Pd/C and 10% Pd/C catalysts.

18

Conclusions

- Using aqueous nutshell extract (*Carya illinoinensis)* and a microwave-assisted synthesis, active AgPd/C and AuPd/C catalysts for the MOR in alkaline media where synthesized.

Table 5. Summary of results for the synthesized and analyzed catalysts

Synthesis Microwave	Catalyst	Metal Loading	Particle size (nm)	Alloying degree (XRD)	Molar Fraction (XRF)	(R_{CT}) Ω
BP – 310	AuPd/C	4.388%	5	$Au_{72}Pd_{28}$	$Au_{46}Pd_{54}$	44742
Discovery CEM	AgPd/C	3.155%	21	$Ag_{27}Pd_{73}$	$Ag_{26}Pd_{74}$	6.8 E11

19

Acknowledgements

- Ph. D. Edgar Valenzuela Mondaca (UABC), collaborator in electrochemical analysis and result discussions.
- M. S. Manuel Román Aguirre (CIMAV), advisor of this work and responsible of physicochemical analysis and result discussions.
- Ph. D. Víctor Hugo Ramos Sánchez, advisor of this work, responsible of physicochemical, electrochemical analysis and result discussions.

20

Thank you!
Any questions?

One-pot one-step synthesis of metal/intermetallic nanoparticles using tandem laser ablation synthesis in solution (LASIS) and chemical reduction method (CRM)

Sheng Hu[1,2], and Dibyendu Mukherjee[1,2,3*]

[1]Sustainable Energy Education and Research Center (SEERC); [2]Department of Chemical&Biomolecular Engineering; [3]Department of Mechanical, Aerospace, & Biomedical Engineering, University of Tennessee, Knoxville

The ability to tune structure-property characteristics of metal/intermetallic nanoparticles (NPs) with unique composition, structure and morphology, via controlled synthesis techniques, can enable physico-chemical, and optoelectronic properties that find wide-spread applications in new classes of energetic, plasmonic, and catalytic materials. To this end, a novel one-step, one-pot solution-phase experimental route is developed that combines *"top-down"* laser ablation synthesis in solution (LASIS) with *"bottom-up"* wet chemical reduction method (CRM). LASIS endows the NPs with metastable crystalline structures and phases due to the extreme liquid-confined plasma conditions, while CRM allows for systematic control of their sizes and shapes. We present a comprehensive study for the synthesis of Co/Co-oxide and Co-oxide/PtNPs using the aforementioned tandem LASIS-CRM technique. The effect of laser fluence, ablation time, temperature and solvent conditions used are investigated. Our results from LASIS on bulk cobalt targets indicate cobaltosic oxide (Co_3O_4) NPs as the main product. In contrast, laser ablation with 532 nm wavelength at laser fluence of ~24.86 J/cm^2 generates cobalt monoxide (CoO) NPs due to thermodynamic conditions arising from different ablation mechanisms. Specifically, Co_3O_4@Pt shell-core NPs are generated from tandem LASIS-CRM in K_2PtCl_4 solution, as confirmed by the elemental mapping from energy dispersive X-ray spectroscopy (EDX) and electron energy loss spectroscopy (EELS) data. Those shell-core nanocomposites, generated via laser assisted galvanic displacement reactions between Co and K_2PtCl_4, exhibit excellent catalytic activity for both oxygen evolution reaction (OER) and oxygen reduction reaction (ORR). Furthermore, Co_3O_4 nanorods (~5-15 nm diameter and ~100 nm length) are synthesized during LASIS at high pH (pH=13-14), which is due to laser breakdown of $Co(OH)_2$ in alkaline conditions. Preliminary electrochemical characterizations for the photocatalytic activities of the aforementioned CoOx nanostructures indicate promising results for their incorporation into future photoelectrochemical cells. The results presented here report the first-ever tandem LASIS-CRM synthesis of controlled nanostructures such as Co_3O_4 nanorods and Co_3O_4@Pt shell-core NPs.

382984 Palladium Nanocrystals for Formic Acid Electro–Oxidation: The Effects of Facet and Twin Defect

Abstract:

Palladium has been extensively studied as a catalyst for the formic acid electro–oxidation (FAO) in direct formic acid fuel cells (DFAFCs) due to its high catalytic ability. Since the FAO activity on single crystal surfaces increases in the order of Pd(110) < Pd(111) < Pd(100), controlling the shape of Pd nanocrystals has been explored with great interest. This presentation introduces a comprehensive study of the FAO on Pd nanocrystals with a variety of different shapes (which preferentially expose different facets) including cubes, right bipyramids (RBPs), octahedra, tetrahedra, decahedra, and icosahedra. The FAO on the various shapes of Pd reveals that the Pd nanocrystals enclosed by {100} facets have higher specific activities than that on the nanocrystals enclosed by {111} facets, with respect to the observations on the Pd single crystals. When comparing nanocrystals which are predominantly enclosed by a particular facet, (e.g., mostly {100} or {111}), the twinned nanocrystals (those having a twin defect) show higher FAO activity than the single nanocrystals. To rationalize these experimental results, we performed density functional theory (DFT–GGA) calculations on model single–crystal surfaces of Pd. The calculations suggest that the enhancement in FAO could be attributed to enhanced flux on Pd(211) sites, a model representation of the twin defects, compared to that on Pd(100) and Pd(111).

Synthesis and Evaluation of High Surface Area Carbon Porous Media for Hydrogen-Bromine Fuel Cell

Venkata Yarlagadda and Trung Van Nguyen
Department of Chemical& Petroleum Engineering
The University of Kansas
Lawrence, KS 66045, USA

Abstract

The active surface area of the carbon electrode plays a significant role in optimizing the performance of a H_2-Br_2 fuel cell. The active surface area of the porous carbon media used as electrodes for the bromine reactions in a H_2-Br_2 fuel cell is quite low. A common approach to increase the active surface area for the bromine reactions is to use multiple layers of these porous carbon media. While this approach increases the active surface area, it also increases the thickness and diffusion pathway for the reactants and products of the reactions and could lead to mass transport limited performance at higher current densities. An alternate approach to increase the active surface area without affecting the transport related morphological properties like porosity and tortuosity of the electrode is to grow nanotubes directly onto the electrode fiber surface to create high active surface area. Figure 1 shows a SEM micrograph of carbon nanotubes on the carbon fibers of a gas diffusion electrode synthesized by our group. This presentation will discuss the synthesis and characterization of this carbon nanotube-based material and its performance as a bromine electrode in a H_2-Br_2 fuel cell

Acknowledgements

This work was funded by NSF through grant number EFRI-1038234 and DOE/ARPA-E under award number DE-AR0000262.

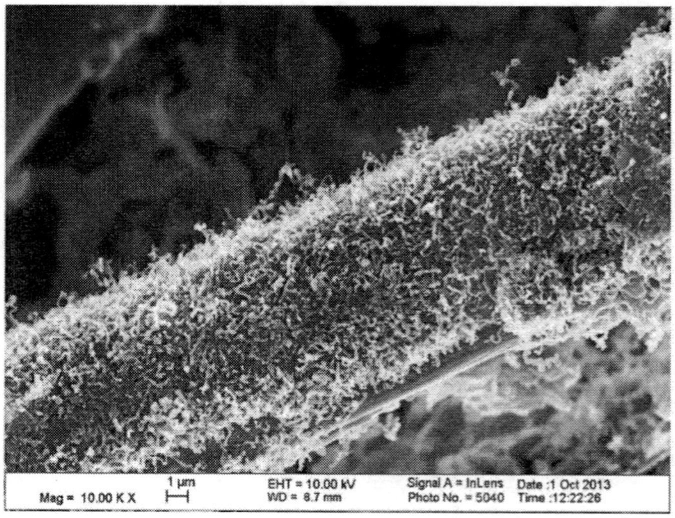

Figure 1. High surface carbon electrode with carbon nanotubes on carbon fibers.

378313 High Performance Non Noble Metal Containing Catalyst for PEM Fuel Cell

Abstract:

High performance Non Noble Metal Containing Catalysts For PEM Fuel Cell

Prasad Patel[a], Moni K. Datta[b], Prashanth Jampani Hanumantha[a], Prashant N. Kumta[a, b, c, d]

[a]*Department of Chemical and Petroleum Engineering* [b]*Department of Bioengineering*

[c]*Department of Mechanical Engineering and Materials Science*

[d]*School of Dental Medicine*

University of Pittsburgh, Pittsburgh, PA 15261

The rapid depletion of fossil fuels and increased environmental pollution due to vast fossil-fuel consumption is a major impetus for efficient use of energy and exploration of renewable and clean energy sources [1-3]. Generation of electricity from renewable energy sources, such as solar, wind without producing carbon dioxide-an undesirable green house pollutant, offers enormous potential for meeting future energy demands [4]. Development of novel and efficient technology to store electrical energy is important for facing the global energy demand and environmental concerns. Fuel cell technology has garnered increasing attention as it provides promising and sustainable approach for the production of continuous power with reduced greenhouse gas emissions and higher efficiencies compared to hitherto combustion based technologies. In particular, proton exchange membrane fuel cells (PEMFCs) are considered to be suitable power sources for automobiles, consumer electronic devices and auxiliary power units due to the advantages of using hydrogen as fuel which is light-weight, clean and has a low operating temperature. Hydrogen also offers quick start-up, extended durability of system components, high power density with low weight and volume due to elimination of additional steps needed for fuel reformation. The simple system design would be reflected as an ease in operation, reduced cost and high reliability. However, capital cost of the system is a major constraint limiting commercialization of PEMFCs due to use of expensive Pt/C catalyst. Hence, development of non-noble metal based catalysts with high electrochemical activity and durability is of interest to replace Pt/C and thus, lower the cost of PEMFC system.

The present study explores $(W_xM_{1-x})O_3$ solid solution as anode electro-catalyst for hydrogen oxidation reaction (HOR) for PEM fuel cell application, where x = 0.7, 0.8. The x-ray mapping shows homogeneous distribution of W, M and O without segregation at any specific site. Electrochemical characterization has been carried out in H_2 saturated 0.5 M sulfuric acid (H_2SO_4) as an electrolyte, Pt wire as counter electrode and Hg/Hg_2SO_4 as the reference electrode (+0.65 V with respect to SHE), using a scan rate of 10 mV/sec and temperature of 40^0C. The solid solution catalyst $(W_{0.7}M_{0.3})O_3$ exhibits superior electrochemical activity and stability, similar to Pt/C. In addition, to obtain a better understanding of the fundamental electrochemical activity and long term stability of $(W_{0.7}M_{0.3})O_3$electro-catalyst, first-principles calculations of the total energies and electronic structures of the model systems with chemical compositions similar to those of the experimentally synthesized materials have been carried out to complement the present experimental study.

Results indicate that $(W_{0.7}M_{0.3})O_3$has the potential for replacing Pt/C, owing to the excellent electrochemical performance and stability. The result obtained to date offers the potential for portending significant reduction in the overall capital costs of PEM fuel cell systems. Results of the structural characterization and electrochemical activity of this novel electro-catalyst will be presented and discussed.

References:

1. Liu, C., et al., *Advanced Materials for Energy Storage.* Advanced Materials, 2010. **22**(8): p. E28-E62.
2. Miller, J.R., *Valuing Reversible Energy Storage.* Science, 2012. **335**(6074): p. 1312-1313.
3. Peng, B. and J. Chen, *Functional materials with high-efficiency energy storage and conversion for batteries and fuel cells.* Coordination Chemistry Reviews, 2009. **253**(23–24): p. 2805-2813.
4. Whittingham, M.S., *Materials challenges facing electrical energy storage.* MRS Bull., 2008. **33**(4): p. 411-419.

373551 Molecular Modeling of Polymer Electrolyte Membrane for Fuel Cell

Abstract:

Molecular Modeling of Polymer Electrolyte Membrane for Fuel Cell

Seung Soon Jang

Computational NanoBio Technology Laboratory, School of Materials Science and Engineering, Georgia Institute of Technology,771 Ferst Drive NW, Atlanta, GA 30332-0245, USA

We present what we have learned from our researches on polymer electrolyte membrane for fuel cell systems using molecular dynamics (MD) simulation method. Mostly our effort has been focused on the relationship between nanophase-segregation and transport properties. From our proton exchange membrane studies, it is clear that the interaction of polymer backbone with water molecules determines the nanophase-segregation, and such nanophase-segregation determines the internal structure of water phase as well as the distribution and dimension of the water phase. Based on these understanding on the proton exchange membranes, we pursue the working principles in anion exchange membranes. For fair comparison, we simulate proton and anion exchange membrane fuel cell using the same polymer backbone with the same molecular variables such as molecular weight, equivalent weight, water contents, and so on. The only difference is the ionic entities: the proton exchange membrane has sulfonate group with hydronium, whereas the anion exchange membrane has trimethyl ammonium group with hydroxide. Quantum mechanical density functional theory (DFT) is used to describe monomeric units with B3LYP and 6-31G**, from which Mulliken population charges are calculated for electrostatic interactions in MD simulations. We use Dreiding force field (FF) for polymers with F3C water model. In addition to the FF for hydronium published in 2004, we prepare the FF for hydroxide. From MD simulations, we investigate the relationship between nanophase-segregation and transport properties in the membranes. The nanophase-segregation is quantitatively evaluated by the structure factor analysis which characterizes the correlation length in reciprocal space. The transport of molecules such as water, hydronium and hydroxide is analyzed using the mean-square displacement.

390114 Enhancing ORR Activity and Durability Via Employing TiO2-Based Supports for Pt Nanoparticles

Abstract:

It is still a challenging issue for the commercialization of the fuel cell to improve the sluggish oxygen reduction reaction (ORR) kinetics and the durability under corrosive ORR condition for cathode electrocatalysts. In this study, we report a promising approach to address these two major issues; the Cr doping to TiO2 supports for Pt-based electrocatalysts. It was clearly revealed that the Cr doping led to a marked enhancement of ORR kinetics which was attributed to the compressive strain in the Pt lattice as well as the increased electronic conductivity of the Cr-TiO_2 supports. Furthermore, Pt/Cr-TiO_2 demonstrated a far superior durability to the conventional Pt/C which was assessed by the accelerated durability tests (ADT) and the *in-situ* XANES studies. While the specific activity for Pt/C was decreased by 43% (141 $\mu A/cm^2$ and 82 $\mu A/cm^2$, for before and after ADT, respectively) after ADT, that for Pt/Cr-TiO_2 was merely reduced by 13% (472 $\mu A/cm^2$ and 409 $\mu A/cm^2$).

379911 Nanoscale Design of Electrocatalysts for Alkaline Fuel Cell and Electrolyzer Applications

Abstract:

Over the last decade, hydrogen has been considered as the key energy carrier, which can be used to convert renewable energy sources via electrochemical energy conversion devices, such as electrolyzers and fuel cells. Utilization of hydrogen depends on two fundamental electrochemical reactions: the hydrogen evolution reaction (HER) in electrolyzers, and its reverse, the hydrogen oxidation reaction (HER) in fuel cells. Although the reaction kinetics of both HER and HOR in alkaline media are significantly slower compared to those in acid electrolytes, the activity trends in alkaline electrolytes have been much less explored. Recent single crystal study has shown that a delicate balance between adsorbed H and adsorbed OH on the bimetallic surfaces can significantly improve both HER and HOR activity in alkaline electrolytes. Thus, the application of the bimetallic mechanisms to design active nanoscale catalysts with sizes of practical relevance is imperative to develop high performance alkaline fuel cells and electrolzyers. In this work, we investigate the HER and HOR in alkaline electrolytes on metal nanoparticles to achieve fundamental understanding of the relationship between surface atomic structure and electrocatalytic activity, using novel synthesis methods and thorough electrochemical characterizations, which aims at developing active nanoscale electrocatalysts.

Effect of Titanium Doping on the Structure and Catalytic Activity of Molybdenum Dioxide for the Reforming of Liquid Hydrocarbons

Qian He[1], Oscar Marin-Flores[1], M. Grant Norton[1], Su Ha[2]

[1] *School of Mechanical and Materials Engineering, Washington State University, Pullman, WA, 99164-2920, USA*
[2] *Voiland School of Chemical Engineering and Bioengineering, Washington State University, Pullman, WA, 99164-2710, USA*

Abstract

Molybdenum dioxide (MoO_2) has demonstrated potential not only as a heterogeneous catalyst but also as electrode material in lithium ion batteries and fuel cells. MoO_2 exhibits high catalytic activity towards the reforming of complex liquid hydrocarbon fuels, such as n-dodecane, jet-A fuel and gasoline. Along with the high catalytic performance, MoO_2 has also displayed high coking resistance and excellent sulfur tolerance, which are key properties in fuel reforming catalysts. The excellent catalytic properties combined with the mixed ionic and electronic conductivity exhibited by MoO_2 have led to the use of this metal oxide as an effective anode material in solid oxide fuel cells. However, stability problems prevent the widespread use of this oxide in commercial applications.

In the present work, MoO_2 was doped with titanium (Ti^{4+}) by using a novel solution-based synthesis method followed by hydrothermal reduction, rather than using a conventional high-temperature solid-state reaction. Nano-sized Ti-doped MoO_2 particles with Ti/Mo ratio of 0, 0.03 and 0.05 were prepared. X-ray photoelectron spectroscopy (XPS) performed on the Ti-doped MoO_2 sample showed that the Ti dopant is present in the Ti^{4+} form. XPS depth profile analysis indicated the presence of Ti^{4+} even in the bulk of the sample, in agreement with the data obtained by using Rietveld refinement on X-ray and neutron powder diffraction patterns showed changes in both the lattice parameters and the Mo-Mo interatomic distance upon doping, suggesting the substitution of Mo^{4+} by Ti^{4+}. Transmission electron microscopy and laser scattering showed that the particle size ranges from 20 to 40 nm. Elemental mapping obtained by scanning electron microscopy with energy dispersive spectroscopy showed homogenous distribution of the dopant.

Attenuated total reflectance-Fourier transform infrared spectroscopy revealed that the solubility limit of the synthesized Ti-doped MoO_2 was reached at Ti/Mo ratio of 0.5, indicating that the bulk thermodynamic solubility limit can be exceeded at the nanoscale.

The Ti-doped MoO_2 sample after annealing in N_2 at elevated temperatures showed the absence of phase changes. High temperature annealing caused increase in the crystallite size and decrease in the surface area most of which comes from the mesoporous structure. X-ray photoelectron spectroscopy analysis suggested the formation of a core-shell structure where the core is formed by the MoO_2 and the shell by the Ti-doped MoO_2. The interaction of Ti-O-Mo was observed by using Fourier transform infrared spectroscopy. Temperature-programmed reduction combined with thermogravimetric analysis carried out on the annealed samples indicated that the presence of Ti^{4+} dopant caused a delay in the onset reduction temperature of MoO_2, which is associated with an increase on the energy of activation barrier imposed by the reduction process. These observations were confirmed using in situ X-ray powder diffraction and led to conclude that Ti-doping improves the stability of MoO_2 when exposed to strong reducing environments. The Ti-doped MoO_2 also showed higher catalytic stability towards the reforming of n-dodecane than undoped MoO_2, which may be related to the decreased oxygen mobility in MoO_2 as a result of the doping.

Related work has been published in Journal of Nanoparticle Research:
Qian He, Oscar Marin-Flores, Shuozhe Hu, Louis Scudiero, Su Ha and M. Grant Norton, J Nanopart Res (2014) 16:2358.

389374 Hierachically Structured Mesoporous Carbons As Fuel Cell Electrode Materials

Abstract:

With a hollow core mesoporous shell geometry, porous carbon can be structured to tune surface area for reaction, nanopore diameter and length for transport, and protection of metal nanoparticles. Studies are reported for methanol oxidation and oxygen reduction using Pt and PtRu nanoparticles supported in a carbon structure composed of hollow core coated with a mesoporous shell (HCMS). The structural parameters can be tuned to study the combined effects of surface area, mesopore size, macropore size, and hierarchy of porosity. The distribution of Pt and PtRu nanoparticles in a mesoporous carbon can vary by tuning synthetic parameters and compared for activity and utilization. For methanol oxidation, the particles dispersed into a thicker shell are more stable and give better performance over time, as shown in voltage cycling and extended methanol oxidation.

Preliminary results on oxygen reduction reaction show that the HCMS structure can offer superior mass transport and a high diffusion limiting current, in addition to durability over cycles of accelerated degradation tests.

379606 Molybdenum Doped Titanium Oxide As a Novel Support for Enhanced Oxygen Reduction Reaction

Abstract:

The search for catalysts with high activity and longer-term stability for ORR in PEMFC is ongoing. In this research, various amounts of molybdenum were doped into anatase TiO_2 and analysed. The electronic conductivity of $Ti_{0.9}Mo_{0.1}O_y$ was significantly improved by doping molybdenum into anatase-TiO_2 structure and hydrogen reduction at 300 ^oC. Later, Pt(111) nanoparticles were reduced and anchored on the thermally treated $Ti_{0.9}Mo_{0.1}O_y$ nanosupports (d- $Ti_{0.9}Mo_{0.1}O_y$). The effects of doping and induced oxygen vacancies were investigated both theoretically and experimentally. Rotating disk electrode (RDE) measurements showed that 20 wt% Pt/d-$Ti_{0.9}Mo_{0.1}O_y$ catalyst had 1.5 times and ~9.1 times higher Pt mass activity for ORR than those of 10 wt% Pt/d-$Ti_{0.9}Mo_{0.1}O_y$ and commercial Pt/C catalysts, respectively. The observed high activity of metal oxide supported-Pt catalyst could be attributed to the role of the advanced oxide support with electron donation from support to Pt catalyst surface, oxygen vacancies on nanosupport surface and high conductivity. Pt nano-electrocatalysts on the advanced robust non-carbon d-$Ti_{0.9}Mo_{0.1}O_y$ nanosupports also exhibits improved stability against Pt sintering under a potential cycling regime (3000 cycles from 0.4 V to 1.0 V vs. RHE) due to Strong Metal-Support Interaction (SMSI). Moreover, the role of oxygen vacancies on oxide surface in enhancing of ORR activity was also elucidated by the computational approach.

366798 Design Principles for Quantum Dot-Based Photocatalysts for Water Splitting

Abstract:

Semiconductor nanocrystals of size in the strong quantum confinement regime, or quantum dots, exhibit size-dependent energy gap. The size-tunable shift of band-edge states can be a powerful tool in the control of exciton dynamics. For example, heterostructure nanocrystals with mere changes in geometric dimensions can lead to the band alignment from type I to type II, which could dramatically increase the exciton recombination lifetime. [1]

In this presentation, we describe the design of heterostructure nanocrystals based on Pb and Cd chalcogenides showing ultralong recombination lifetime. For the study, we synthesized PbSe/CdSe core/shell nanocrystals and introduced CdS outer-most shell with varying thickness and morphology. Spectroscopic study revealed that as the outer CdS shell becomes thicker, the quantum confinement in the shell region is relieved and thus the heterostructure has quasi-type-II energy alignment. Furthermore, photoluminescence lifetime of the sample also increased when the CdS shell became thicker: nearly 2-orders-of-magnitude increase was observed. Notably, slight change in CdS deposition condition resulted in varying morphology of the CdS shell. Not only could we design heterostructure nanocrystals with enhanced electron-hole separation, the structure could turn into "open" geometry, such as tetrapods. In the reduction of methylene blue, we observed noticeable increase in photocatalytic activity as the geometry became open.[2] For instance, tetrapod-shaped PbSe/CdSe/CdS core/shell/shell nanocrystals exhibited enhanced photocatalysis compared to pyramid-shaped PbSe/CdSe/CdS nanocrystals, let alone spherical structures. Our study of photocatalysis using PbSe-based heterostructure nanocrystals suggests that the geometric nature of nanocrystals should be considered in the context of the efficient conversion of photon energy into chemical reactions. [3]

We extended our design principle into CdSe quantum dot / TiO_2 composite photocatalysts. The new photocatalytic system shows controlled and improved photocatalytic efficiency of water splitting. Overall water splitting using heterostructure nanocrystals will be discussed.

References

[1] Lee D C, Robel I, Pietryga J M and Klimov V I 2010 *J. Am. Chem. Soc.* **132** 9960

[2] Pak C, Woo J Y, Lee K, Kim W D, Yoo Y and Lee D C 2012 *J. Phys. Chem. C* **116** 25407

[3] Kim W D et al. 2014 Chem. Commun. **50** 1719

Morphological Modification of TiO_2 films by Polyethyleneglycol (PEG) and Their Photoelectrochemical Performance in the Presence of Glycerol as a Hole Scavenger

Muhammad Ibadurrohman, Klaus Hellgardt

PEG-modified TiO_2 films (xPEG-TiO_2, x is the concentration of PEG 200 in the precursor solution in g/L) were successfully synthesised by a simple, versatile, and low-cost spray pyrolysis deposition method. The effects of PEG 200 as a surface modifier on the relevant properties of TiO_2 films were investigated by means of SEM (surface morphology), XRD (crystallography), stylus profiler (thickness), UV-Vis absorbance spectroscopy (optical responses and band gap), and photoelectrochemical measurements (photoelectrolytic performance as photoanodes). The pure TiO_2 films show regular, dense and compact nanoparticle network with relatively large grain size (mostly above 500 nm), well-defined grain boundaries, and a film thickness of ca. 1090 nm. Adding an amount of 10 g/L of PEG 200 into the precursor spraying solution results in smaller grain size (mostly below 400 nm) and thinner film (860 nm). Increasing the concentration of PEG 200 in the solution further decreases the grain size and film thickness, yet increase the roughness of the films (Figure 1a-e). The decrease of film thickness upon PEG addition may be attributed to the viscosity increase of the solution due to PEG introduction, which, at the same operating nozzle pressure, leads to lower spraying flow rates. XRD measurements of the pure TiO_2 film indicate that the TiO_2 crystal phase is of anatase with an observed characteristic peak at the 2θ of 25.3° which corresponds to (1 0 1) plane. This characteristic peak gradually disappears upon increasing the concentration of PEG 200. UV-vis absorbance measurements reveal that the band gap of pure TiO_2 film is typically 3.19 eV and slightly increase to 3.20, 3.25, 3.28, and 3.30 eV for the case of 10PEG-TiO_2, 20PEG-TiO_2, 30PEG-TiO_2, and 40PEG-TiO_2, respectively.

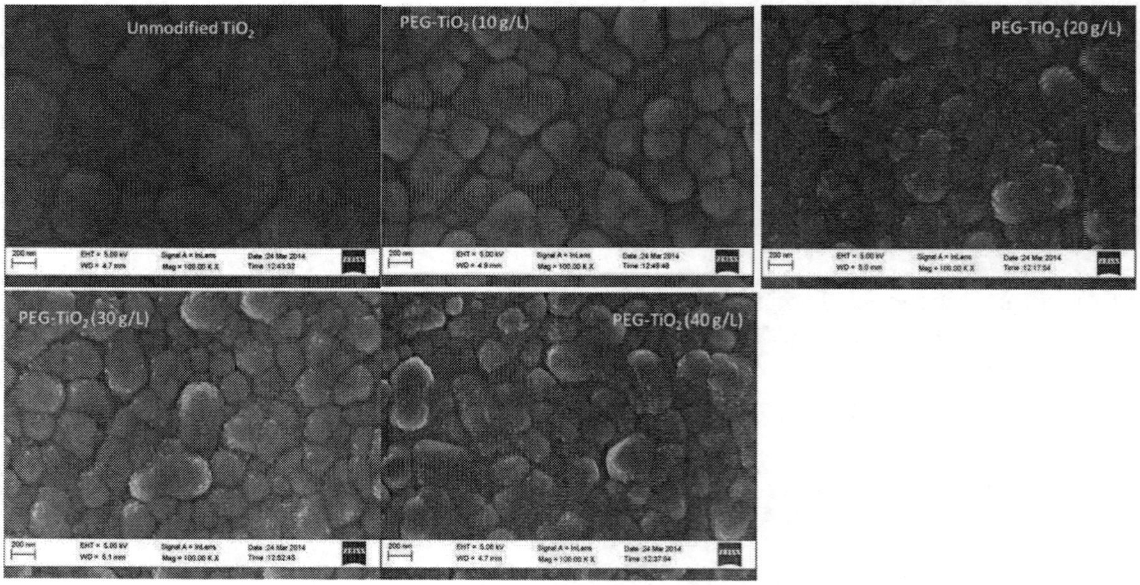

Figure 1. SEM images of pure TiO_2 (a), 10PEG-TiO_2 (b), 20PEG-TiO_2 (c), 30PEG-TiO_2 (d), 40PEG-TiO_2 (e)

Bias-dependant photoconversion efficiency, which is calculated from Linear Sweep Voltammetry (LSV) measurements in 1M NaOH electrolyte, gives a maximum value of 0.49% at a relatively small bias (-0.35 V vs HgO|Hg) in the use of unmodified TiO_2 photoanode. Interestingly, the use of 10PEG-TiO_2 results in a significant improvement in maximum photoconversion efficiency, giving a value of 1.25% at an even more negative bias (-0.46 V HgO|Hg). Higher content of PEG 200 leads to a monotonous decrease of photoconversion efficiency to

0.99% at -0.41 V vs HgO|Hg, 0.83% at -0.44 V vs HgO|Hg, and 0.68% at -0.42 V vs HgO|Hg in the case of 20PEG-TiO_2, 30PEG-TiO_2, and 40PEG-TiO_2, respectively (Figure 2a). In order to further confirm the significance of PEG 200 as a surface modifier for TiO_2 films, we prepare pure TiO_2 films with different thicknesses by varying the number of layers during spray pyrolysis deposition processes, resulting in pure TiO_2 films with thicknesses ranging from 546 to 1090 nm. These films give photoconversion efficiency of 0.33 – 0.49%, much lower than the PEG-modified TiO_2 films with a similar thickness range (Figure 2b). The effectiveness of the PEC cell employing 10PEG-TiO_2 as an excellent photoanode was further examined by running it in the presence of glycerol as a hole scavenger. Glycerol plays an important role as an effective sink for the photogenerated holes so that the charge carrier recombination can be significantly supressed and, subsequently, the photocurrent is enhanced (Figure 2c). The additional photocurrent due to glycerol introduction into the PEC cell strongly depends upon the initial concentration of glycerol according to a model resembling Langmuir-Hinshelwood isotherm (Figure 2d).

The outcome of this research is expected to be an initial step towards green production of energy (electricity and H_2) from renewable sources (water and sunlight) using a vastly available semiconducting material (TiO_2) as the photoanode, polyethyleneglycol as the surface modifier, and industrial waste (glycerol) as the hole scavenger.

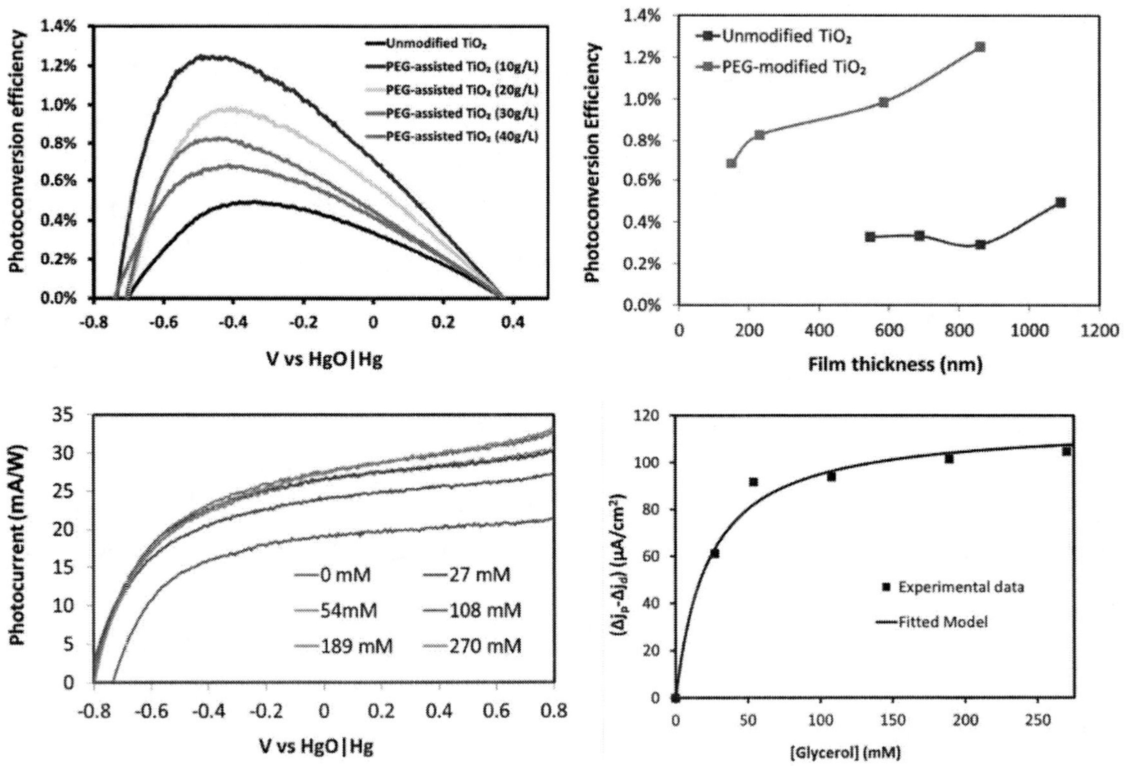

Figure 2. Bias-dependant photoconversion efficienct of as-prepared films calculated from j-V curves (a), photoconversion efficiency comparing unmodified TiO_2 and PEG-modified TiO_2 as a function of film thickness (b), j-V curves of 10PEG-TiO_2 in the presence of glycerol as a sacrificial agent (c), additional photocurrent due to glycerol introduction as a function of its concentration, data fitted with Langmuir-Hinshelwood adsorption model (d)

Photoelectrochemical Water Reduction of a Copper(I) Oxide Photocathode Made Using An Inkjet Printing Approach

ABSTRACT

Sunlight assisted Photoelectrochemical (PEC) water splitting is of great interest to researchers for producing clean hydrogen and hydrocarbons (by reducing carbon dioxide) in order to meet the growing global energy demand for renewable energy resources. The biggest obstacle to successful implementation of PEC water splitting is the lack of availability of a semiconductor with suitable band gap and band edge energetics and durability.[1] To maximize photon absorption while providing sufficient energy to facilitate reduction of carbon dioxide and oxidation of water, band gap values for photocatalysts are required in the range of 2 to 2.5 eV.[2-3] Cu_2O is a promising semiconducting material for photoelectrochemical water splitting due to its band gap (E_g = 2.0 eV) and its band edge positions straddling the water oxidation and reduction reactions. The major drawbacks with Cu_2O include its poor aqueous stability, short carrier diffusion length (20–100 nm), high absorption depth (10 microns) and difficulty in synthesis and film processing.[2] The present work targets three challenges: (1) providing a simple commercializable process for the synthesis of a stable Cu(I) oxide ink, (2) fabrication of high quality copper (I) oxide (free from Cu and CuO impurities) films using inkjet printing for the first time, and (3) the fundamental study of various protective nanolayers of ZnO, TiO_2 and WO_3 on the photoelectrochemical water reduction process.

REFERENCES

1. Paracchino, A.; Laporte, V.; Sivula, K.; Graetzel, M.; Thimsen, E., Highly active oxide photocathode for photoelectrochemical water reduction, *Nature Materials,* **2011**, *10,* 456-461.
2. Martinez-Garcia, A.; Vendra, V. K.; Sunkara, S.; Haldankar, P.; Jasinski, J.; Sunkara, M. K., Tungsten oxide-coated copper oxide nanowire arrays for enhanced activity and durability with photoelectrochemical water splitting, *J. Mater. Chem. A,* **2013**, *1*, 15235-15241.
3. Liao, P.; Carter, E. A., New concepts and modeling strategies to design and evaluate photo-electro-catalysts based on transition metal oxides, *Chem Soc Rev,* **2013**, *42*, 2401-2422.

Tightly adhered Ta_3N_5 nanotube arrays as a photoanode for solar water splitting

Peng Zhang, Tuo Wang and Jinlong Gong*

Key Laboratory for Green Chemical Technology of Ministry of Education, School of Chemical Engineering and Technology, Tianjin University; Collaborative Innovation Center of Chemical Science and Engineering, Tianjin 300072, China
E-mail: jlgong@tju.edu.cn

H_2 generation via solar water splitting is considered to be one of the most promising solutions to the energy crises, which takes the advantages of the inexhaustibility, universality and environmental benignancy of solar energy.[1] Among different semicondutor photocatalysts, Ta_3N_5 is thought to be an outstanding candidate, which can untilze visible-light and realize unassisted soalr water splitting owning to its proper band gap energy (~ 2.1 eV) and band structure.[2] The theoretical photoconversion efficiency of Ta_3N_5 is as high as 15.9% under AM1.5 irradiation.[3]

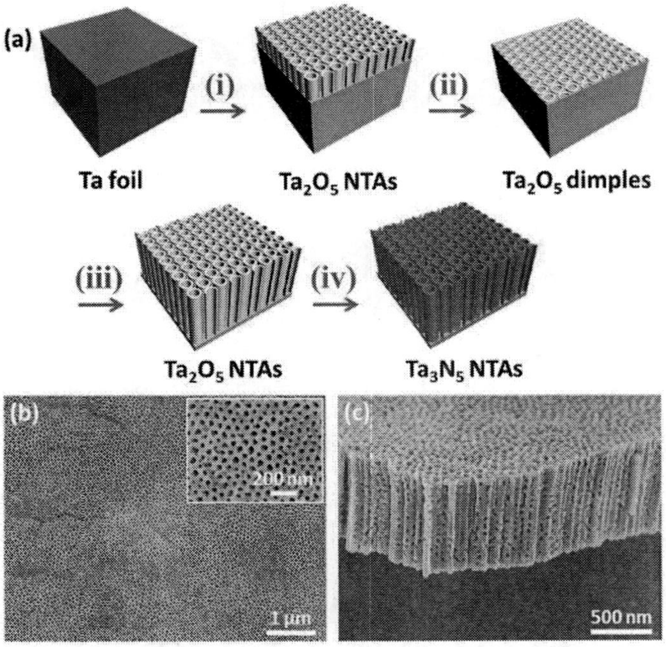

Figure 1. (a) Schematic illustration of the synthesis process of Ta_3N_5 NTAs: (i) first anodization step, (ii) sonicaton, (iii) second anodization step and (iv) nitridation. (b) Top-view SEM image of Ta_3N_5 NTAs (inset shows a high-magnification top-view SEM image). (c) Cross-sectional SEM image of Ta_3N_5 NTAs.

In this work, fine Ta_3N_5 nanotube arrays (NTAs) were synthesized by combining an improved two-step anodization method with the subsequent high-temperature (900 °C) nitridation (Fig. 1a). The obtained Ta_3N_5 NTAs showed a highly ordered structure and a tight adhesion with the Ta substrate (Fig. 1b, c). It is hypothesized that the tight adhesion could improve the transport efficiency of the photogenerated electrons to the cathode.

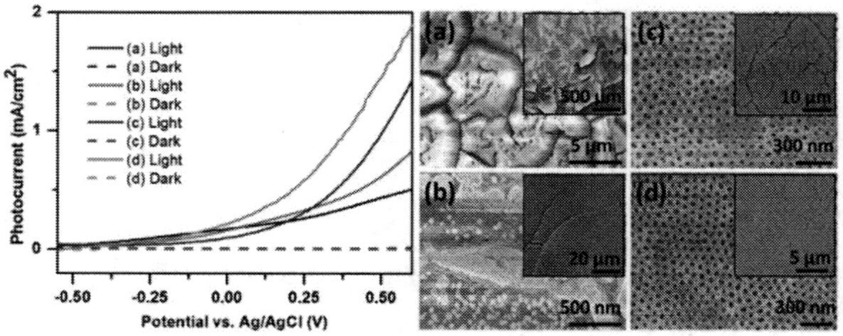

Figure 2. I-V curves of Ta_3N_5 NTAs after nitridation of different Ta_2O_5 NTAs (with top-view SEM images, respectively) synthesized in various conditions: (a) one-step anodization in the solution consisted of 27 ml H_2SO_4, 0.5 ml HF and 2.5 ml H_2O at room temperature; (b) one-step anodization in the solution consisted of 27 ml H_2SO_4, 0.3 ml HF and 2.7 ml H_2O at room temperature; (c) two-step anodization in the solution consisted of 27 ml H_2SO_4, 0.3 ml HF and 2.7 ml H_2O at room temperature; (d) two-step anodization in the solution consisted of 27 ml H_2SO_4, 0.3 ml HF and 2.7 ml H_2O at 0 °C. The anodizations were conducted with a bias of 15 V for 10 min.

In the anodization step, the low concentration of HF in the electrolyte helped prevent the etching of the interface between Ta_2O_5 material and Ta substrate, resulting in a better adhesion and improved photocatalytic properties (Figure 2a, b). The two-step approach led to the formation of Ta_2O_5 nanotubes with well-defined structure and ordered surface topography, providing more inner surface of the nanotubes for photocatalytic reactions together with an enhanced photocurrent density (Figure 2b, c). The ice-water bath during the anodization was also critical to the rapid heat removal and the decrease of reaction rate, which improved the adhesion of the NTAs and the photoelectrochemical properties (Figure 2c, d). The results supported the hypothesis that the photocatalytic properties of Ta_3N_5 NTAs is correlated with their adhesion force to the substrate. A stronger attachement of Ta_3N_5 NTAs will result in a more conductive transport pathway for photogenerated electrons and a higher photocatalytic activity.[4]

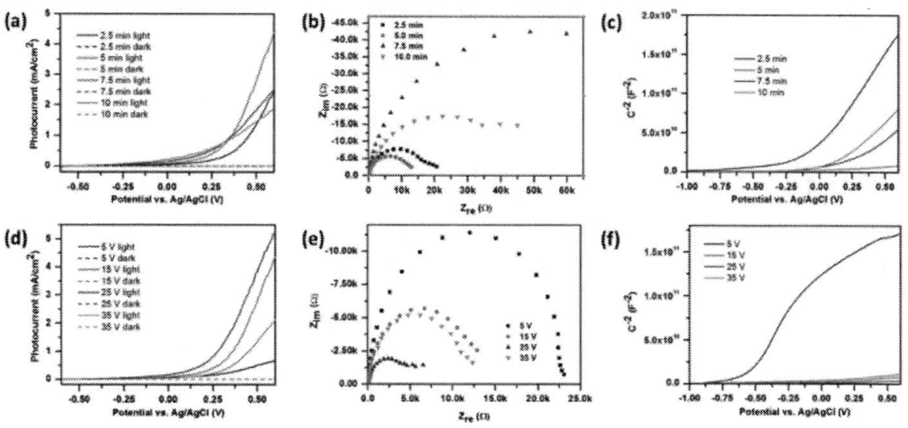

Figure 3. (a) I-V curves; (b) EIS (in Nquist plot) and (c) Mott-Schottky plots of Ta_3N_5 NTAs synthesized with different anodization durations. The anodization bias is 15 V for all samples. (d) I-V curves; (e) EIS (in Nquist plot) and (f) Mott-Schottky plots of Ta_3N_5 NTAs synthesized with different anodization bias. The anodization duration is 5 min for all samples.

The anodization duration and bias were also tuned to further prove the correlation between the adhesion force of Ta_3N_5 NTAs and their photocatalytic activities for solar water splitting. Ta_3N_5 NTAs with the optimized anodization duration of 5 min and bias of 25 V showed the highest photocurrent of 5.3 mA/cm^2 at 0.6 V *vs.* Ag/AgCl (1 M NaOH, pH 13.6) (Figure 3a, d). The photoelectrochemical properties of Ta_3N_5 NTAs obtained with different anodization durations and biases were closely related to their strength of adhesion (as proved by the EIS studies and Mott-Schottky plots shown in Figure 2b, c, e and f). When the anodization was conducted for a short duration, the tubular structure was barely formed. However, longer anodization duration would lead to larger volume expansion during the material transformation from Ta foil to Ta_2O_5 and the peeling off of the Ta_3N_5 NTAs. The optimized anodization bias would provide a proper migration rate of metal ions and the oxidizing species (O^{2-} or OH^- ions in the electrolyte) so that the reaction would processed in a much moderate manner to form well-attached Ta_3N_5 NTAs.

In summary, we have employed an improved anodization method to synthesize tightly adhered Ta_3N_5 NTAs photoanode. The two-step approach improves the surface topography and regularity of the NTAs. The effects of low HF concenteration and decreased temperature in the electrolyte help control the rate of the anodization reaction thus enhance the adhesion of the NTAs. With the stronger connection between the NTAs and the substrate, the transport of photogenerated electrons to the counter electrode could be improved. The photoelectrochemical performance of the Ta_3N_5

NTAs was further improved by investigating the anodization durations and the biases. The highest photocurrent of 5.3 mA cm^{-2} at 0.6 V *vs.* Ag/AgCl (1 M NaOH, pH 13.6) was obtained for the sample synthesized with a bias of 25 V and a duration of 5 min. The correlation between adhesion force of NTAs and their photoelectrochemical performance was fruther proved by EIS data analysis and Mott-Schottky plots, which suggested that enhancing the semicondutor film-substrate adhersion force was an effective way to improve the activity of photoanode in photoelectrochemical solar water splitting systems.

References:

[1] a) M. G. Walter, E. L. Warren, J. R. McKone, S. W. Boettcher, Q. Mi, E. A. Santori, N. S. Lewis, Chem. Rev. 2010, 110, 6446; b) A. Kudo, Y. Miseki, Chem. Soc. Rev. 2009, 38, 253.

[2] P. Zhang, J. Zhang, J. Gong, Chem. Soc. Rev. 2014, DOI: 10.1039/C3CS60438A.

[3] A. Murphy, P. Barnes, L. Randeniya, I. Plumb, I. Grey, M. Horne, J. Glasscock, Int. J. Hydrogen Energ. 2006, 31, 1999.

[4] a) Y. Li, T. Takata, D. Cha, K. Takanabe, T. Minegishi, J. Kubota, K. Domen, Adv. Mater. 2013, 25, 125; b) Y. Li, L. Zhang, A. Torres-Pardo, J. M. González-Calbet, Y. Ma, P. Oleynikov, O. Terasaki, S. Asahina, M. Shima, D. Cha, L. Zhao, K. Takanabe, J. Kubota, K. Domen, Nat. Commun. 2013, 4.

380879 ZnO Supported Ultrathin Au Nanowires for Photoelectrocatalytic Applications

Abstract:

Fuel cells are believed to be promising technology to challenge the increasing energy demand particularly direct alcohol oxidation reaction displays high energy density. This work presents the synthesis of hybrid ZnO/ultrathin Au nanowires[1] and their support and geometry-dependent electrocatalytic performances towards alcohol oxidation upon illumination. Geometry control of Au nanostructures is a useful method to modify its electrocatalytic properties– by tuning electrochemical active surface area, exposed facets and aspect ratio.[2] Grown ultrathin Au nanowires exhibited outstanding electrochemical activity that is much higher than that exhibited by Au nanoparticles. ZnO support provides an excellent stability for Au nanowires during reaction and also produced reactive charge carriers under excitation, thus assisting in oxidation current generation. Since both ZnO and Au nanowires exhibited ethanol oxidation reaction independently via photoirradiation and applied potential, respectively, the direct coupling of Au nanowires and ZnO, hybrid displayed impressive photoassisted electrocatalytic performance and excellent stability.[3]The mechanism of the synergetic effect of ZnO supported Au nanowires towards photoelectrocatalytic activity is proposed based on electrochemical measurements. We envision that the as-synthesized semiconductor-ultrathin nanowires hybrids hold great possibility for integration as catalyst in future fuel cell technology.

References

1. Halder, A.; Ravishankar, N., Ultrafine Single Crystalline Gold Nanowire Arrays by Oriented Attachment. *Advanced Materials* **2007,***19*(14), 1854-1858.

2. Porter, N. S.; Wu, H.; Quan, Z.; Fang, J., Shape-control and electrocatalytic activity-enhancement of Pt-based bimetallic nanocrystals. *Accounts of chemical research* **2013,***46*(8), 1867-1877.

3. Drew, K.; Girishkumar, G.; Vinodgopal, K.; Kamat, P. V., Boosting fuel cell performance with a semiconductor photocatalyst: TiO_2/Pt-Ru hybrid catalyst for methanol oxidation. *The Journal of Physical Chemistry B* **2005,***109* (24), 11851-11857.

364837 Virus-Templated Nanostructured Titania-Lead Sulfide Heterojunction Solar Cells

Abstract:

Current titania and quantum dot solar cells are generally structured in a bilayer. However, a bicontinuous nanostructured morphology promotes more efficient exciton dissociation and charge transport. Such nanostructured composites are difficult to obtain without the use of energy intensive processes to create regular structures, and often nanopores cannot be obtained. The M13 bacteriophage, which is easy and inexpensive to amplify and has the ability to template a variety of metallic and oxide materials, can be organized into a nanoporous film via layer-by-layer assembly. Here, we use the bacteriophage films as a scaffold to generate a titania nanomesh and infiltrate it with lead sulfide quantum dots. The porosity of the virus-templated titania film can be tuned as a function of the titania synthesis time. These parameters, as well as the nature and infiltration method of lead sulfide quantum dots, affect the final heterojunction solar cell performance. In addition, the continuous pathways created in the virus-templated titania film reduce recombination compared to traditional titania paste films. In summary, we demonstrate that M13 bacteriophages represent a promising tool to organize nanomaterials and create continuous nanostructures that can lead to efficient photovoltaic devices.

Photonic Curing of Ligand-Capped $CuInSe_2$ Nanocrystals

C. Jackson Stolle, Taylor B. Harvey, Brian A. Korgel

McKetta Department of Chemical Engineering, Center for Nano- and Molecular Science, Texas Materials Institute, The University of Texas at Austin, Austin, TX 78712

Semiconductor nanocrystal inks are being explored as a route to make low-cost, high efficiency photovoltaic (PV) devices. Nanocrystals can be synthesized in large quantities, dispersed in solvents, and deposited under ambient conditions on virtually any substrate. PV devices made from $CuInSe_2$ nanocrystals have achieved efficiencies of 3% without any high-temperature processing, however the efficiency is limited due to poor interparticle charge transport. Higher efficiencies have been achieved through high temperature sintering of $Cu(In,Ga)Se_2$ nanocrystals in a selenium rich atmosphere (known as selenization). This processing step is difficult to control and adds significant manufacturing cost. We have explored photonic curing as a means to eliminate the need for high temperature selenization and also achieve high efficiency from $CuInSe_2$ nanocrystal devices. Photonic curing utilizes a high-intensity pulsed light to transiently heat the nanocrystal film, allowing for nanocrystal sintering on temperature-sensitive substrates such as plastic. Photonic curing is also capable of high-throughput roll-to-roll manufacturing.

We find that the capping ligand chemistry on the nanocrystal surface plays a large role in film morphology after photonic curing. $CuInSe_2$ nanocrystals are synthesized with oleylamine capping ligands, which provide stable dispersions in organic solvents. We find that during photonic curing, high pulse energies are required and the oleylamine-capped nanocrystals sinter through a melting and resolidification mechanism which leaves significant regions of exposed back conact. Replacing oleylamine with inorganic chalcogenidometallate clusters (ChaMs) has been shown to improve nanocrystal film electrical properties and also aids in the sintering of $CuInSe_2$ nanocrystal films. We find that, when treated with photonic curing, ChaM-capped $CuInSe_2$ nanocrystals sinter at lower pulse energies without the films undergoing melting and resolidification. This leads to substantially improved film morphology.

370984 Selenization of Automated, Ultra-Sonic Spray-Deposited Cu(In,Ga)Se2 Nanocrystal Films for Photovoltaics

Abstract:

Direct conversion of sunlight to electricity using photovoltaics (PVs) has the potential to be a source of energy that is both cost effective and clean. Lower PV manufacturing cost is needed to realize this promise. One possible route to lower cost is to eliminate vacuum processing during the absorber layer deposition using copper indium gallium selenide (CIGS) nanocrystal inks. Efficiencies of 7.1% and 12% have been reached by annealing films of $CuInGaSe_2$ and $GuInGaS_2$ nanocrystals in a selenium atmosphere, also known as selenization. While promising, these results were achieved with small area devices using deposition techniques that are potentially scalable, but will require significant modification.

As the next step in developing this technology, we have used a highly scalable, automated, ultra-sonic spray coating system. Utilizing this technology leads to finer control in the fabrication process; however, significant variations were observed in film morphology. These variations were traced to small changes in the organic content of the nanocrystal ink. The organic content of the film was found to be highly dependent on the amount of anti-solvent used during the nanocrystal washing procedure. An anti-solvent/solvent centrifugation wash procedure is used in almost every nanocrystal synthesis, but the importance of this procedure on device performance has been overlooked. We trace small changes in the washing of the nanocrystals to morphology changes that then cause dramatic changes in device performance.

By implementing tighter controls on the entire nanocrystal PV device fabrication, we have demonstrated reproducible 5% efficiencies using a scalable deposition technique. We also have explored the effect of air pressure, nanocrystal concentration, and liquid flow rate on the ultrasonic spray deposition of nanocrystal solar inks.

Abstract Title:
Polymer Electrolyte Integration via iCVD for Applications in Dye Sensitized Solar Cells

Author:
Yuriy Y. Smolin, Steven Farrell, Sruthi Janakiraman, Masoud Soroush, and Kenneth K.S. Lau*

Keywords:
Polymer Electrolyte, Dye Sensitized Solar Cell, Initiated Chemical Vapor Deposition, poly(4-vinylpyridine), poly(vinylpyrrolidone), pore filling

Section:
Nanomaterials for Photovoltaics (T5000)

Abstract:
Initiated chemical vapor deposition (iCVD) is used for the synthesis and integration of polymer electrolyte thin films, such as poly(4-vinylpyridine) (P4VP) and polyvinylpyrrolidone (PVP), within the mesoporous TiO_2 photoanode of dye sensitized solar cells (DSSCs). DSSCs with the conventional liquid-electrolyte are prone to leakage and evaporation—hindering DSSC applications, durability, and thermal stability. Polymer electrolytes do not suffer from these disadvantages and can also enhance the cell current-voltage (I-V) behavior. However, spin coating and drop casting to deposit the polymer electrolytes lead to incomplete pore filling of the polymer inside the mesoporous TiO_2 photoanode, resulting in poor electrical contact and lower efficiency. To alleviate these pore filling concerns, the polymer electrolytes were fabricated using the novel, solvent-free technique of initiated chemical vapor deposition (iCVD)[1,2]. iCVD is generally an adsorption limited technique where the reagents for polymerization, the monomer and initiator, are heated to a vapor that can easily penetrate into the mesoscale voids of the TiO_2 photoanode. The initiator is activated by a heated filament (250-350 C) and polymerization as thin films occurs inside small crevices and on surfaces.

By understanding the competition between surface reaction kinetics and mass transport into the mesovoids, iCVD polymerization allows for uniform coatings on nanoscale geometries and is an exceptional technique for allowing polymerization within the mesoporous structure of TiO_2. The surface reaction kinetics and mass transport have been found to depend on the relative pressure of monomer, $z = \frac{P_M}{P_M^{sat}}$, which is an adsorption parameter that provides a measure of the surface availability of the monomer. By carefully controlling z, nearly 100% pore filling of the TiO_2 structure has been achieved as estimated by thermogravimetric analysis (TGA)[3]. This is significantly higher than that achievable with liquid techniques like spin coating or drop casting of polymer solutions. This work will show that iCVD polymer electrolytes can be effectively integrated within TiO_2 mesoporous photoanodes to produce enhanced DSSCs. By varying the polymer electrolyte chemistry, DSSC characteristics, including open circuit voltage, short circuit current density and fill factor, can be tuned. To gain a better understanding on the effect of the polymer electrolyte, experimental techniques such as UV-VIS absorption, intensity modulated spectroscopy, and impedance spectroscopy have been used. Furthermore, FTIR, NMR, and XPS spectroscopic analyses confirm that the polymers formed are stoichiometric in composition expected of linear homopolymers and chemically identical to ones formed with liquid solution methods.

* Corresponding Author: klau@drexel.edu

(1) Lau, K. K. S.; Gleason, K. K. *Thin Solid Films* **2008**, *516*, 678.
(2) Lau, K. K. S.; Gleason, K. K. *Macromolecules* **2006**, *39*, 3688.
(3) Nejati, S.; Lau, K. K. S. *Nano Letters* **2010**, *11*, 419.

Diffusion Mechanism of Na in Molybdenum Thin Films for Improving $Cu(In,Ga)Se_2$ Solar Cell Efficiency

Robert Forest, Erten Eser, Brian McCandless, Robert Birkmire, Jingguang Chen

$Cu(In,Ga)Se_2$ (CIGS) is a promising semiconductor material for manufacturing low cost photovoltaics. The presence of sodium (Na) in CIGS is well known to increase the open-circuit voltage and solar cell efficiency, but this effect is not well understood. Na can be added to CIGS from a variety of sources, but the simplest method is to allow Na to diffuse out of a soda-lime glass (SLG) substrate through a layer of molybdenum (Mo), which serves as an electrical contact to the solar cell. The downside of this technique is that it can result in spatially non-uniform cells over larger areas, although it is still favored due to its simplicity and low cost. To better understand how Na from soda-lime glass is incorporated into CIGS, this presentation investigates the mechanism for the transport of Na from a SLG substrate through the Mo electrical contact.

The surface accumulation technique was used to study the diffusion of Na in Mo films. Mo coated SLG samples were heated in vacuum at various temperatures and the concentration of Na at the surface was measured as a function of heating time using X-ray photoelectron spectroscopy (XPS). The observed Na accumulation was consistent with Harrison type C kinetics which occurs solely through grain boundaries. The Mo accumulation data was fit using the solution to the transport equation developed by Hwang and Balluffi for Type C diffusion. The grain boundary diffusion coefficient was estimated at several temperatures and found to increases with temperature following an Arrhenius relationship. The apparent activation energy ~~was~~ lower than expected for physical grain boundary diffusion but is similar to the dissociation of the Mo-O bond. Additionally, it was found that samples with a greater amount of oxygen on the surface of the Mo accumulated more Na. These results suggest that oxygen residing in the grain boundaries and on the surface of the Mo film is involved in the diffusion of Na through Mo.

377999 Role of Redox Electrolyte on the Performance of CdSe Based Quantum Dot Solar Cell

Abstract:

Inorganic quantum dots with their well defined properties such as size-tunable band gaps, high luminescence quantum efficiency, and excellent photostability, have been successfully used as sensitizers in the photoelectrochemical solar cells. The efficiency in a such device is determined by the intricate balance achieved between the various charge separation and recombination reactions. Large differences in the timescales of electron injection (femto- second) and transport (milli second) processes, amplifies the rate of recombination. Improvement in the performance requires precise estimation and tuning of rate constants for all the charge injection and recombination processes. Here, we report the charge injection and recombination dynamics of CdSe quantum dots anchored on TiO_2 and SiO_2 in the presence of various liquid electrolytes and its influence on the light conversion efficiency. The ability to alter the energy gap between CdSe and TiO_2 as well as CdSe and redox couple has profound influence on the interfacial charge transfer kinetics. Femto second absorption spectroscopy and photoluminescence lifetime measurements have been carried out to monitor the charge injection kinetics in various electrolytes as a function of redox couple with different redox potential, concentration and pH of the medium. Various strategies to maximize photoinduced charge separation and electron transfer processes for improving the overall efficiency of light energy conversion will be discussed.

Summary (full abstract next page):

With solution-process-ability, scale-able fabrication and purification, and cheap input materials, semiconducting single-walled carbon nanotube (SWNT) networks represent promising materials for near-IR solar cell (SC) applications. This promise has motivated a body of work not only developing solar cells[1-4] but also exploring alignment/deposition methods and SWNT photovoltaic (PV) physics. Despite this interest, there is to date *no* quantitative model of SWNT solar cell operation analogous to bulk semiconductor p-n junction PVs, allowing a rigorous understanding of the physical tradeoffs driving experimental observations and informing what research will enable technological progress. Even the most basic considerations of different chemical (e.g., chirality mixture) and structural (e.g., nanotube alignment, density, film thickness, etc.) properties remain unknown. In this work we have derived the steady state operation of planar heterojunction SWNT PVs from the fundamental light absorption, exciton transport, and free carrier transport behaviors of single nanotubes. Our method can treat arbitrary distributions of nanotube chiralities, lengths, orientations, defect types and concentration, bundle fraction and size, density, dielectric environment, electrode combinations, etc. We achieve this by treating individual SWNT properties as random variables, and describing the network by the dependent distributions of those properties, yielding coupled stochastic differential equations for light absorption, exciton transport, and free carrier transport. Applying the model to monochiral (6,5) films in aligned and isotropic configurations, we find that there is a strongly optimal film thickness at a given nanotube network density and orientation, reflecting an inherent tradeoff between light absorption (i.e. exciton generation) and diffusion to the electrodes. This optimal shifts lower with increasing density, and is ultra-thin (<10 nm) for horizontally-aligned films but 50-200 nm for vertically aligned films. We further reveal that this trend is perfectly captured by a power-law relationship with constants that are invariant with quenching site concentration. We show that due to weak inter-SWNT exciton transport relative to exceptional intra-SWNT diffusion, vertically-aligned films are unambiguously favored at densities above 3% of close-packed; at lower densities however an optimum emerges at an intermediate angle to compensate for weaker light absorption of vertical nanotubes. Films with in-plane aligned nanotubes are unambiguously the worst design. Our predictions are quantitatively consistent with to-date experimental devices.

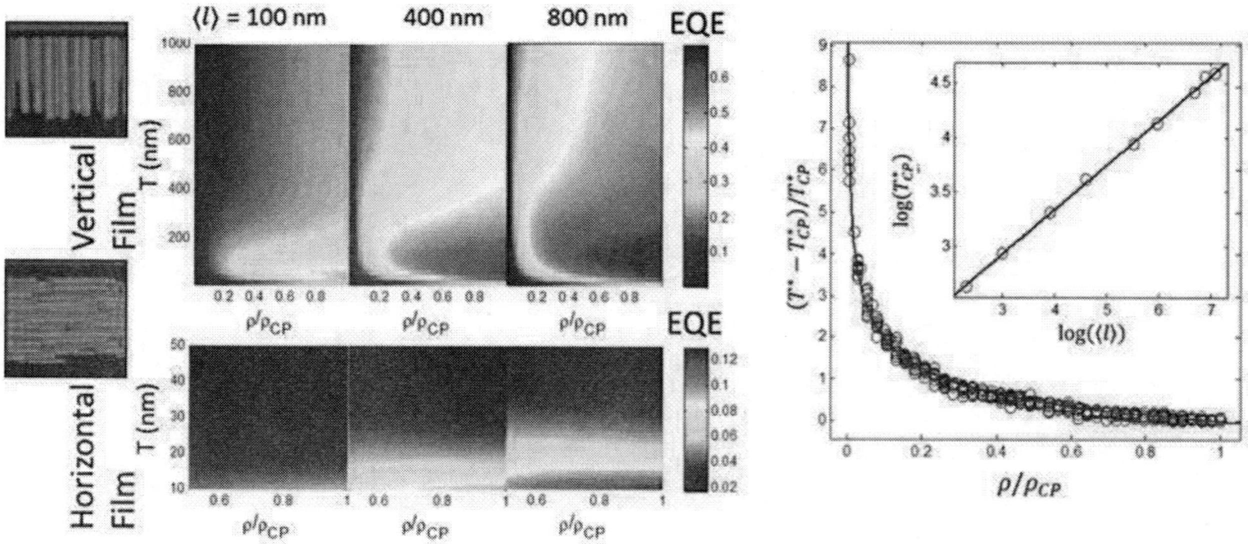

Full Abstract:

Both cost and performance attributes make semiconducting single walled carbon nanotubes (SWNT) promising photo-absorbers for near-infrared photovoltaic (nIR PV) applications.[1-7] Their solution-process-ability (which profoundly reduces manufacturing costs), earth-abundant source material (carbon), and recently scale-able fabrication and purification can make SWNT Solar Cells (SCs) cheap compared to conventional SCs.[4-6, 8-10] Furthermore, such devices can augment the photo-conversion efficiencies of conventional visible-PV systems by absorbing nIR wavelengths that comprise approximately 22% of the solar spectrum and fall within typical PV bandgaps; low unit-operations costs particularly enable this application by making stacked independent PV layers (four-terminal SCs) feasible, versus the constraints of one-step fabrication (*i.e.* lattice, bandgap, and workfunction matching constraints) that have challenged tandem multi-junction (two-terminal) SCs.[11] SWNT are also attractive relative to many other organic and inorganic PV materials through exceptional exciton and free carrier diffusivity and mobility, exceptional absorption cross-sections, and resistance to oxidation by water and oxygen.[4-6, 12-15] To date, single-walled nanotube (SWNT) SCs with 0.01-1.7% external quantum efficiency (EQE) have been developed employing single chiralities,[2, 3] mixed semiconducting chiralities,[1, 4] and mixed semiconducting and metal chiralities[5, 6] in bulk heterojunction[4, 7] and planar heterojunction[1-3] configurations using polymers[16] and fullerenes[1-4] for exciton dissociation and/or co-photoabsorption.

Despite this progress, there exists no model describing steady-state SWNT PV operation for the selection of design variables such as SWNT chirality, density, layer thickness, nanotube orientation, *etc.* Even the most basic questions of SWNT SC design cannot be answered: should a mixture of chiralities be employed to maximize absorption across the solar spectrum, or is a single-chirality film preferable to minimize exciton trapping at low-bandgap nanotubes? Is a vertically aligned film favored to maximize exciton transport to electrodes, or a horizontally aligned film to maximize light absorption, or an intermediate balance between the two? What thicknesses and densities – even to an order of magnitude – are optimal? How important are nanotube length, the presence of impurities, and alignment? By analogy, quasi-fermi-level transport models and the depletion zone approximation describing bulk semiconductor p-n junction PVs have steered research and enabled technology development in conventional SCs. Hence, advancement of SWNT nIR PVs demands a quantitatively predictive framework permitting comparison of chemical (chirality mixture, defect types and concentrations, *etc.*) and geometric (absolute and relative alignment of SWNT, bundle fraction, *etc.*) influences on performance. Recognizing this need, qualitative arguments are being made on SWNT SC structural and chemical optimization by considering intuitive trade-offs between light absorption and charge collection, but the lack of quantitative comparative methods has led to dissenting opinions on *e.g.* the chirality mixture and orientation angle questions above.[6]

In this work we have discerned an approach, the first, to capture deterministically the efficiency of a SWNT nIR PV derived only from the fundamental physics of SWNT light-matter interaction, exciton transport, and free carrier transport. We do so by treating relevant local (single-SWNT) properties as random variables, and describing the network of nanotubes as a distribution of those properties. We focus specifically on planar heterojunction devices. This treatment rests on a simplifying approximation of macroscopic ($O(100\ nm)$) homogeneity. Our method applies to arbitrary networks in terms of chemical and structural properties – *e.g.* any distribution of nanotube chiralities, lengths, orientations, defect types and levels, bundle fraction and size, density, dielectric environment, electrode combination, *etc.*

Rigorously, it is limited to densities above the percolation threshold, although it has quantitatively reproduced recent observations at low densities in literature.

Applying it to monochiral (6,5) films, our model yields a wealth of predictions that clearly indicate the direction experimental investigations should take. Films exhibit an optimal thickness that varies with alignment, density, and impurity content as a consequence of balancing light absorption (better at higher thicknesses) and exciton transport to electrodes (better at lower thicknesses). The alignment angle (*e.g.* vertical, horizontal, intermediate, isotropic) impact is a result of longitudinal (along-nanotube) exciton diffusivity being five orders of magnitude or more greater than orthogonal (inter-nanotube network) diffusivity; competing with that effect, light absorption is, across the solar spectrum, approximately six times greater along the nanotube longitudinal axis than along the axial direction, leading to lower light absorption but higher exciton diffusivity for vertically-aligned films. At sufficiently high densities (above 3% of close-packed) vertically aligned films are unambiguously favored to maximize exciton transport, but as density lowers below 10% of close-packed a second optimum emerges for some thicknesses at an intermediate angle – $\pi/3$ to $\pi/4$ from the vertical axis – to compensate for the decreasing light absorption. We also reveal the relative importances of density, alignment, length distribution, and impurity concentration. Of all film properties, density and alignment are revealed to play dominant roles in determining efficiency, with 10- to 100-fold performance improvements achieved through vertical alignment and increasing density above 10% of close-packed ($O(10^{17}\ m/m^3)$ length of SWNT per volume). The implications of these predictions are that next generation devices should focus on higher densities and vertical alignment, and thicknesses in the range of 40-100 nm depending on their other properties.

1. Bindl, DJ, Shea, MJ, et al., *Chem Phys*, 2013. **413**: p. 29-34.
2. Jain, RM, Howden, R, et al., *Adv Mater*, 2012. **24**(32): p. 4436-4439.
3. Shea, MJ and Arnold, MS, *Applied Physics Letters*, 2013. **102**(24): p. 243101.
4. Bernardi, M, Lohrman, J, et al., *Acs Nano*, 2012. **6**(10): p. 8896-8903.
5. Wang, L, Liu, H, et al., *Chemical Society Reviews*, 2013. **42**(20): p. 8134-8156.
6. Arnold, MS, Blackburn, J, et al., *Physical Chemistry Chemical Physics*, 2013.
7. Ye, Y, Bindl, DJ, et al., *Small*, 2014: p. n/a-n/a.
8. Liu, H, Nishide, D, et al., *Nat Commun*, 2011. **2**: p. 309.
9. Tvrdy, K, Jain, RM, et al., *Acs Nano*, 2013. **7**(2): p. 1779-1789.
10. Moshammer, K, Hennrich, F, et al., *Nano Res.*, 2009. **2**(8): p. 599-606.
11. Brown, AS and Green, MA, *Progress in Photovoltaics: Research and Applications*, 2002. **10**(5): p. 299-307.
12. Schoppler, F, Mann, C, et al., *J Phys Chem C*, 2011. **115**(30): p. 14682-14686.
13. Naumov, AV, Tsyboulski, DA, et al., *Chem Phys*, (0).
14. Ruzicka, BA, Wang, R, et al., *Phys Rev B*, 2012. **86**(20).
15. Gabor, NM, Zhong, ZH, et al., *Phys Rev Lett*, 2012. **108**(8).
16. Ren, SQ, Bernardi, M, et al., *Nano Lett*, 2011. **11**(12): p. 5316-5321.

382277 Doped TiO_2 Based Core-Shell Structures for High Efficiency Hybrid Solar Cells

Abstract:

Hybrid solar cells, with an inorganic/organic interface for charge separation, have been extensively investigated in the past decade in order to replace the expensive Si based technology with an inexpensive alternative. Typically, these devices incorporate a mesoporous TiO_2 film which is decorated with dye molecules and filled with a hole transport polymer, for example P3HT, to conduct the electrons and holes, respectively. Recently, we have shown that the efficiency of nanowire based hybrid solar cells can be increased from ~1.8 % to 2.5 % through the formation of a Sn-doped $TiO_2|TiO_2$ core-shell device created via a hydrothermal growth and subsequent $TiCl_4$ treatment. However, this surface treatment presents difficulties in creating a crystalline conformal coating, limiting the control over the extent of coating and the crystallinity, directly affecting the charge injection from the polymer into the TiO2 array. In this work, we directly deposit a controllable TiO_2 film through atomic layer deposition to conformally coat the nanowire arrays with various thicknesses. By changing the thickness and TiO_2 crystallinity, we are able to engineer the energy levels at the TiO_2-dye-P3HT interface due to the magnitude and position of the Fermi levels of the core and shell material, influencing the rate of charge injection and recombination. Furthermore, the crystallinity of the shell layer directly affects the amount of dye that can be absorbed on the surface of the nanostructures with a reduction in light absorption by roughly 30% from anatase to rutile TiO_2. Finally, a detailed mechanism will be proposed for the device performances based on the energy level alignment between the pinned Fermi-level TiO_2 structure and the HOMO of the P3HT resulting in a shifting open circuit voltage based on the crystal phases. Additionally, the core-shell structures are characterized with photovoltage decay and impedance spectroscopy measurements to study the charge transport and recombination across these various interfaces.

Magnetic Assembly of a Nickel Capped Zinc Oxide Nanorod Network for Use as DSSC Working Electrode

Roger Chang, Kemakorn Ithisuphalap and Ilona Kretzschmar
Department of Chemical Engineering, The City College of New York,
City University of New York, 140th Street and Convent Avenue, New York, NY 10031

Dye-sensitized solar cells (DSSC) are a promising technology for power generation that is both renewable and does not produce climate changing emissions. Improvements in DSSC performance have been achieved through advances in dye and electrolyte chemistry and more recently, through the use of nanofabrication techniques. By changing the typical nanoparticle morphology of the semiconductor in a DSSC working electrode, efficiency improvements have been attained.

Zinc oxide can be used as a viable alternative to titanium dioxide in a DSSC working electrode due to the fact that the two semiconductor materials have nearly identical band gaps. In addition, zinc oxide is easily grown into a variety of morphologies and zinc oxide exhibits much higher electron mobility than titanium dioxide (115 to 155 $cm^2/V\text{-}s$ versus 10^{-5} $cm^2/V\text{-}s$ respectively). Researchers have taken advantage of these properties to build DSSCs based on zinc oxide nanowire array working electrodes. However, their efficiencies have been limited by an inadequate surface area for dye loading. The motivation for our work is to take advantage of zinc oxide's high electron mobility in 1-D structures while also allowing for a higher surface area. By creating a networked structure based on 1-D nanorods, we are attempting to find an optimal solution that will improve the performance of zinc oxide DSSCs.

In our work, nickel capped zinc oxide nanorods are fabricated by the alternating electrodeposition of nickel and zinc ions into the pores of a track-etched porous polycarbonate template. Due to the high porosity of the templates (10^9 pores/cm^2), large numbers of nanorods can be produced efficiently and economically during each deposition. After isolating them from the polycarbonate, the free-floating nanorods are exposed to a magnetic field. Because of the magnetic properties of nickel, the nickel caps become magnetized and it is believed that they can be used to guide the assembly of a large interconnected semiconductor network that is capable of efficient electron transport.

In our experiments, DSSCs fabricated with *unordered* zinc oxide nanorods (aspect ratio of 5:1) have already shown performance improvements over DSSCs fabricated with zinc oxide nanoparticles. The maximum current density improved from 50 $\mu A/cm^2$ to nearly 90 $\mu A/cm^2$ and was attributed to more efficient transport of electrons in a nanorod versus a nanoparticle. Achieving an *ordered* network of nanorods is expected to improve DSSC performance to an even greater extent.

385442 Use of Mesoporous Titania Thin Films with Controlled Pore Orientation in Photovoltaic Applications

Abstract:

Mesoporous titania thin films with 2D Hexagonal Close Packed (HCP) cylindrical nanopores have been synthesized by an evaporation-induced self-assembly (EISA) technique with Pluronic surfactant F127 (with average structure $(EO)_{106}(PO)_{70}(EO)_{106}$ where EO is an ethylene oxide unit and PO is a propylene oxide unit) as the template. To provide vertical alignment of the pores, surface modification of substrates with crosslinked surfactant F127 has been used to provide a chemically neutral surface. Conditions are found giving vertically aligned, 2D HCP titania films with pores near 8-9 nm in diameter, which are precisely the structures expected to provide short carrier diffusion length and high hole conductivity required for an efficient bulk heterojunction solar cell. Anatase titania is a n-type semiconductor with the valence and conduction bands located at +3.1 and -0.1 eV relative to the Fermi level, thus giving a band gap of +3.2 eV. Therefore, titania readily absorbs UV light with a wavelength below 387 nm. Because of this, these titania films can be used as a window layer with a p-type semiconductor incorporated into the pores and at the top surface of the device to synthesize a photovoltaic cell. The pores provide opportunities to increase the surface area for contact between the two semiconductors, to align organic semiconductors, and to induce quantum confinement effects.

Here, these titania films with hexagonal phase are infiltrated with a hole conducting polymer, P3HT, in order to create a p-n junction for an efficient hybrid solar cell, by spin coating followed by thermal annealing. This assembly is hypothesized to give better photovoltaic performance as compared to a disordered or bicontinuous cubic nanopore arrangements (K.M. Coakley et al. *MRS Bull.* **2005** *30*, 57). Confinement in cylindrical nanopores is expected to provide isolated, regioregular "wires" of conjugated polymers with tunable optoelectronic properties, which have been shown to improve hole conductivity over that in bicontinuous cubic structure [M.D. McGehee, *MRS Bull.* **2009**, *34*, 95]. The kinetics of infiltration into the pores show that maximum infiltration occurs within less than one hour in these films, and give materials with improved photovoltaic performance relative to planar TiO_2 / P3HT assemblies. These oritneted mesoporous titania films are also used to develop an inorganic solar cell by depositing CdTe at the top using the Close Spaced Sublimation (CSS) technique. Conversion efficiencies of at least 5-7% are observed in photovoltaics prepared using mesoporous titania films, which are significantly enhanced relative to planar TiO_2/CdTe devices. These mesoporous titania films have a great potential in inorganic solar cell development and can potentially replace CdS window layer which is conventionally used in inorganic CdS-CdTe solar cells.

Integration of Photosystem I Proteins into Solid-State Solar Cells

Maxwell Robinson, Patrick Wellborn, Siyuan Yang, Evan Gizzie, David Cliffel, and G. Kane Jennings

Photosystem I (PSI) is a large membrane protein complex found within the thylakoid membranes of plants and some cyanobacteria, converting incident sunlight into charge separation for the synthesis of sugar. Much research focus has been placed on incorporating extracted PSI proteins as films into a variety of biohybrid devices for solar energy conversion. Thus far, these efforts have been limited to wet-cell techniques, relying on aqueous charge mediation between the P_{700}^{+} and F_B^{-} redox sites to electrode surfaces; order-of-magnitude improvements in photocurrents have been observed through specific tuning of the substrate, PSI film thickness, mediator species and concentration, as well as electrode materials. This presentation will cover recent efforts to develop solid-state biohybrid solar cells incorporating PSI films, both pure and composite, between contacting electrodes. In solid-state systems, mediation is controlled by PSI-PSI charge transfer or via incorporated conductive solid species. Our group has focused on pairing PSI with a variety of complementary components including emeraldine polyaniline, graphene oxide, and reduced graphene oxide to achieve high-performance composite films that can be used in either wet cells or solid-state configurations. This presentation will focus on the integration of PSI with doped variants of poly(3,4-ethylenedioxythiophene) (PEDOT) to improve the performance and stability of solid-state PSI films. Additionally, this presentation will detail the unique fabrication of solid-state PSI solar cells using 100-nm thick nanoporous gold sheets as a light-transparent top electrode.

Engineered Porous Silicon Counter Electrodes for High Efficiency Dye Sensitized Solar Cells

William R. Erwin †, Landon Oakes ‡, Shahana Chatterjee §, Holly F. Zarick †, Cary L. Pint §, and Rizia Bardhan †*

†Department of Chemical and Biomolecular Engineering, ‡Interdisciplinary Materials Science Program, §Department of Mechanical Engineering, Vanderbilt University, Nashville, TN 37235, USA

Dye-sensitized solar cells (DSSCs) have emerged as a promising third generation solar cell due to efficiencies so far measured up to 12% and straightforward processes for solar cell manufacturing. In a conventional DSSC platinum is used as the counter electrode due to its corrosion resistance to the iodide/triodide redox species utilized as electrolyte, and its excellent electrocatalytic activity toward the I^-/I_3^- redox couple. However the use of this precious metal ($50/gram) remains a barrier to scalability and commercialization of DSSCs providing an incentive to develop DSSC counter electrodes using cheaper, sustainable materials.

In this work, we show a novel DSSC architecture with carbon coated porous silicon as a counter electrode. We have demonstrated the passivation of highly reactive porous silicon with graphenic carbon and use of this material as a counter electrode exhibiting an efficiency of 5.38%, which is comparable to the 5.80% efficiency of a cell made with the conventional platinized counter electrode. We have studied the effect of chemical vapor deposition carbonization parameters on coverage and quality of graphene coating, and shown how this affects overall cell performance. An atomically thin coating of few layer graphene was found to be optimal for passivation of porous silicon as well as activation of the surface for triiodide reduction. We used electrochemical (cyclic voltammetry, electrochemical impedance spectroscopy) methods, spectroscopic (Raman) techniques, and electron microscopy to fully characterize the carbonized porous silicon, and evaluate the performance of the material as a reduction catalyst for the I^-/I_3^- reaction at the DSSC cathode.

389572 CVD-Fabricated Organic Active Layer and Hole Transport Material for Organic-Inorganic Mesoscopic Solar Cell

Abstract:

Unsubstituted polythiophene (PTh) was synthesized using a unique oxidative chemical vapor deposition (oCVD) approach. To enable oxidative polymerization of thiophene monomer, antimony pentachloride was used as an efficient oxidant. The deposited polymer was simultaneously doped to its conductive form. Upon dedoping, the polymer was found to absorb strongly at 510 nm. The optical band gap of the sample was determined to be around 1.9 eV and its LUMO orbital level was estimated to be 2.8 eV. Poly(3,4-ethylenedioxythiophene) (PEDOT) was synthesized by oCVD in the same way with antimony pentachloride as the oxidant. The deposited, doped film electrical conductivity was found to be non-linearly dependent on the inverse of polymer film thickness. Conductivity as high as ~1000 S/cm was estimated from sheet resistivity measured using 1 mA source current. The high conductivity of the film in the as deposited state is attributed to the presence of antimony complexes as dopant within the film, as exchanging these species with different ions including chloride and sulfate resulted in inferior conductivity. X-ray diffraction indicates that the PEDOT film is non-crystalline. The antimony complexes in the film were characterized using Auger electron and x-ray photoelectron spectroscopy. The antimony oxidation state was found to be +3. This along with the Raman spectroscopy results suggest that the PEDOT is in an over-oxidized state which could be the main reason behind the near metallic conductivity of the films. The conductivity in air and at room conditions (21 °C and ~30% RH) was monitored over 30 days and the loss in conductivity was found to be minimal. The oxidant complexes in the film were the main reason for the persistent conductivity, with a Sb/S ratio of ~ 1 yielding the most stable film conductivity for the period of testing. To fabricate an all solid-state organic-inorganic solar cell, a titania mesoporous electrode was deposited on a 50 nm compact layer of titanium dioxide which served as the electron transport layer. A thin layer of polythiophene (equivalent to ~5-8 nm thickness) was then deposited within the sample using oCVD and further dedoped by condensing methanol vapor within the film followed by annealing at 90 °C. The results were compared with the case where the samples were taken out and washed with methanol to remove all antimony complexes from the sample. We have found that the complete removal of antimony is essential for device performance. This can be attributed to the presence of antimony species at the interface which is known to be a strong electron scavenger. Upon drying, the samples were coated with PEDOT as a hole transport layer by oCVD until ~80 nm overlayer of PEDOT was achieved on the assembly. The assembled devices with a gold contact on the top were tested and a 0.8 % photoconversion efficiency without any further optimization was achieved.

Development of Platinum/Graphene Counter Electrode Based DSSCs

Yinghe Zhang[a], Shigeki Kuroiwa[a], Tawfique Hasan[b]

[a] Applied Chemistry, Waseda University, Japan
[b] Cambridge Graphene Centre, Cambridge University, UK

Dye-sensitized solar cells (DSSCs) are an attractive alternative to conventional silicon-based solar cells, largely due to its simple fabrication process and low-cost materials. However, the cost of platinum counter electrode and its dissolution in the I^{3-}/I^{-} electrolyte are drawbacks of the well-established platinum counter electrode based DSSCs. Graphene is a promising candidate for platinum substitution or coating due to the high surface area, low charge transfer resistance and corrosion resistance. In this work, graphene ink is produced by liquid phase exfoliation of pristine graphite. The Pt counter electrode and graphene ink coated on Pt counter electrode are made by various coating methods. N719 is used as the dye in our devices. The coatings are characterized by scanning electron microscopy and Raman spectroscopy. Performance of the devices is compared under a solar simulator. We find that graphene coated Pt counter electrode based DSSCs show efficiency as high as 1.25±0.10%, with 0.71 V open circuit voltage, 0.358 fill factor and 210.36±14.38 Ω series resistance. The device efficiency is 6 times (0.19±0.05 %) higher than that of pure Pt electrode using the same device structure. Our results indicate graphene ink have the potential to be a viable, low cost counter electrode material for DSSCs.

References:

1. Sanedrin, R. G., D. G. Georganopoulou, S. Park, C. A. Mirkin, Adv. Mater., 2005, 17, 1027

390374 Electrochemical Characterization of Photosystem I (PS I)/Self-Assembled Monolayer (SAM)/Au Substrates: The Critical Bottle-Necks in Electron Transfer

Abstract:

Photosystem I (PS I), a supra-molecular protein complex, functions as a biological photodiode that charge separates upon exposure to light and is responsible for driving natural photosynthesis. The highly efficient photo-electrochemical activities of PSI (100% quantum efficiency over 54% of the solar spectrum) make it an ideal candidate for next-generation bio-hybrid electronic devices. Our specific interest in using PS I for future hybrid photovoltaic (PV) device fabrications requires optimal encapsulation of these proteins onto chemically grafted SAM substrates. Our previous results indicate that various experimental parameters alter the surface attachment dynamics of PS I, deposited from colloidal aqueous buffer suspensions onto OH-terminated alkanethiolate/Au SAM substrates, thereby resulting in complex structural arrangements which affect the electron transfer and capture pathway of PSI. We present surface topographical and electrochemical characterizations of PS I/Au SAM substrates to elucidate photo-activated electronic activities of PS I monolayer. Specifically, light induced directional electron transfer by surface immobilized PS I is demonstrated via direct electrochemistry measurements from PS I complexes assembled on alkanethiolate SAM/Au surfaces prepared with varying thiol chain lengths (C2, C4, C6, C8). The results reveal the effect of interfacial thiol brush density and chain length on limiting the electron transfer rate at the donor side from Au surface to oxidized P700+ reaction center in PS I, due to non-optimal packing and/or, large tunneling distance through the SAM. On the other hand, at the acceptor side, the electron donation rates from reduced FB- sites into the solution is significantly increased with the addition of a soluble electron acceptor, Methyl Viologen (MV2+

previous work shone light on the critical role of oxygen in the electron uptake process, both via direct mediation from FB-

O2. To this end, the present work takes a step towards detailed electrochemical characterizations that elucidate and optimize the true bottleneck of electron mediation through PS I monolayer by separately analyzing the interfacial electron transfer from SAM substrate to P700+ and from FB- to the solution.

) in the presence of dissolved oxygen. Our

to O2 and through a secondary complex formation between MV2+ and

366108 Polymer Fabric-Based Thermoelectric Devices

Abstract:

Most thermoelectric devices on the market today are rigid solid-state inorganic based, which possess many drawbacks including high manufacture cost, restriction in installation, and concerns over thermal expansion stress. We have developed a new material system in whcih we can adjust the carrier type and concentration in the solution phase growth. The material can be coated on conventional fabrics, such as nylon, and exhibits high electrical conductivity and reasonable Seebeck coefficient even without thermal annealing. We also demonstrate that the nanomaterial-coated fabrics, indeed, for the first time, can generate considerable amount of electrical power from a small temeparture difference.

Metal Organic Heat Carrier Nanofluids

Satish K. Nune, Peter B. McGrail, Andy Zwoster, Jeromy Jenks, Paul F. Martin, Liem X. Dang and Harsha VR Annapureddy

Nanofluids, dispersions of metal or oxide nanoparticles in a base working fluid, are being intensively studied due to improvements they offer in thermal properties of the working fluid. However, the proven impacts on thermal conductivity are modest and well described from long established effective medium theory. We present a new class of metal-organic heat carrier (MOHC) nanofluid that offers potential for a larger performance boost in thermal vapor-liquid compression cycles. MOHCs are nanophase molecular engineered particles to reversibly uptake and release selected organic compounds used as working fluids in an organic Rankine cycle (ORC) for geothermal power production and waste heat recovery. An ORC system utilizing the MOHCs, therefore, extracts more heat from the geothermal source for a given mass flow rate concomitant with delivery of higher inlet enthalpy of the working fluid to the turbine. We will present our recent results on large-scale synthesis of uniform nanophase MOHC in water and their performance testing in methanol and 1,1,1,3,3-Pentafluoropropane (HFC-245fa) and heat load effects of various MOHC nanofluids (Figure 1).

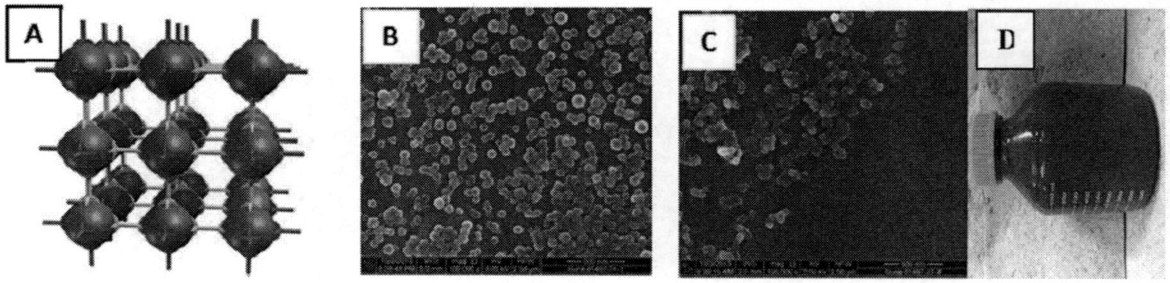

Figure1: (A) Illustration of nanophase MOHC, (B and C) SEM image of nMOHC-1 and nMOHC-2 and (D) permanently stable dispersion of 1 wt % 1 liter nano MOHC.

377391 Facile Synthetic Approach for PbTe and $CuFeS_2$ Nanocrystals and Their Preliminary Thermoelectric Device Setup

Abstract:

Recent years have witnessed increasing public concern on environmental issues caused by industrial operations[1] and unsustainable consumption of fossil fuel[2]. Thermoelectric (TE) materials are capable of converting waste heat into electricity, and are emerging as a promising alternative source of energy. When the dimensions of the material are reduced to nanometer scale, quantum confinement effects are introduced, which can alter the electronic structure[3], resulting in an improvement of thermoelectric efficiency[4].

In this presentation, altaite phase PbTe[5] and chalcopyrite phase $CuFeS_2$[6] were synthesized using a facile solution-phase method. The average diameters of as-synthesized PbTe and $CuFeS_2$ nanocrystals are 13 ± 3 nm and 6.4 ± 0.5 nm, respectively. At the same time, both of them show outstanding thermoelectric properties (ZT of PbTe nanocrystals can reach 0.9, while ZT of $CuFeS_2$ is 77 times the value of bulk chalcopyrite). The success of functional thermoelectric devices through a simple low-temperature solution coating of nanocrystals on glass fibers suggests the possibility of combining the high performance with other attractive properties including flexibility, lightweight and shock resistance.

[1] P. M. Vitousek, H. A. Mooney, J. Lubcheno and J. M. Melillo, Science, 1997, 277, 494-499.

[2] M. Gratzel, Nature, 2001, 414, 338.

[3] L. D. Hicks, T. C. Harman, X. Sun and M. S. Dresselhaus, Phys. Rev. B: Condens. Matter., 1996, 53, R10493.

[4] A. I. Boukai, Y. Bunimovich, J. Tahir-Kheli, J. K. Yu, W. A. Goddard III and J. R. Heath, Nature, 2008, 451, 168-171.

[5] D. Liang, H. Yang, S. W. Finefrock and Y. Wu, Nano Lett., 2012, 12, 2140-2145.

[6] D. Liang, R. Ma, S. Jiao, G. Pang and S. Feng, Nanoscale, 2012, 4, 6265-6268.

386927 Interface Engineering for High Performance Thermo-Electric Nanocomposites

Abstract:

Thermoelectric devices convert thermal energy directly into electrical energy and vice-versa and hold promising contributions in waste heat recovery and refrigeration. Due to their low cost, potential for high throughput manufacturing and unique transport mechanisms; solution-processed conducting polymer/nanoparticle composite films provide an environmentally clean and efficient route to harness electricity from low-grade heat sources. However, we still lack a fundamental understanding of these hybrid systems that would provide a general and robust framework to guide materials design. In this work, we use a model organic/inorganic hybrid system to demonstrate how engineering of nanoscale interfaces can drive novel macroscale transport behavior.

Here, we study a hybrid organic/inorganic thermoelectric model system comprising of Poly(3,4 ethylenedioxythiophene):poly(styrenesulfonate) (PEDOT:PSS) and Tellurium nanowires as a function of nanowire loading and chemical doping. Interesting non-monotonic electrical conductivity is observed at intermediate loadings, suggesting non-effective medium behavior, in contrast to the values of the Seebeck coefficient that increase linearly with mass fraction of inorganic. These trends in Seebeck coefficient and electrical conductivity lead to an optimized power factor in the composites that exceed that of the individual components. The surprising electrical conductivity behavior can be explained with a model where carrier transport is primarily through a highly conductive volume of polymer that exists at the nanoparticle-polymer interface. Additionally, by doping the individual components separately, we can boost the electrical conductivities even further and obtain higher power factors. Thus, a mix of traditional doping mechanisms and innovative interface engineering at the nanoscale allows us to generate high performance thermoelectric materials.

In a complementary study, modifying the interface of hybrid polyvinylpyrollidone (PVP) and Tellurium nanowires by controlled addition of small molecules, allows us to tune the nature of carrier transport from hole-dominated to electron-dominated in thin film devices of these materials. To create a thermoelectric module, complementary hole (p-type) and electron (n-type) transporting materials are necessary, connected electrically in series and thermally in parallel. Typically, one needs two different materials with matching conductance which presents an engineering challenge. However, our approach provides a n-type thermoelectric material from essentially a p-type material simply by modifying the interface while preserving the conductivity and still rendering it solution processable. This enables use of a single material for both legs of the thermoelectric module. The approach is general and we can use it to tune the transport behavior in other promising thermoelectric materials too.

In summary, we provide novel design routes which highlight the role of the interface for optimizing performance in nanoscale organic/inorganic composites and provide guidelines for improving efficiencies of existing devices even further and observing novel transport behavior as well.

390141 Fabrication of Bulk Thermoelectrics from Large-Scale Assemblies of Zn_3P_2 and ZnO Nanowires

Abstract:

Thermoelectrics, solid-state heat engines, are both useful for the recovery of waste heat or the generation of renewable energy. The lack of any moving parts, portability, very long operational lifetimes, and minimal maintenance requirements make them very attractive for these applications. Current state-of-the-art thermoelectric modules not only exhibit low efficiencies, but are also composed of toxic and rare-earth elements. The lack of a pathway for precisely and independently tuning the thermal and electrical transport through materials employed in the fabrication of thermoelectrics is the primary reason for their low efficiencies. Recent theoretical and experimental studies have indicated that nanostructuring of materials serves as a route for independently tuning the thermal and electrical transport through them, and enhance their heat-to-electricity conversion efficiencies. Of all the forms of nanomaterials, nanowires hold the most promise for the fabrication of efficient thermoelectric modules. The two different dimensions of the nanowires, namely the diameters and the lengths, offer a convenient path for independently tuning the electrical and thermal transport through them. As thermoelectrics is a bulk application, large-scale deployment and use of nanowire-based thermoelectric modules requires strategies for the synthesis and assembly of nanowire powders on an industrial scale. In this context, this work describes strategies for the bulk production of Zn_3P_2 nanowire powders and their *in-situ* functionalization with conjugated organic molecules (1,4-benzenedithiol (BDT) and 4-aminothiophenol (4-ATP)), together with the strategies useful for assembling them in an interface-engineered manner. The thermoelectric performance of these inorganic-organic hybrids will be discussed. Also, the optimization of the microstructure and composition of the nanowire assemblies for realizing a 500 fold increase in the thermoelectric performance to a *ZT* value of 0.23 will be discussed. Optimal doping of ZnO nanowire assemblies for obtaining a *ZT* value of 0.6 will also be discussed.

AUTHOR INDEX

AUTHOR INDEX

AUTHOR INDEX

AIChE
120 Wall Street, FL 23
New York, NY 10005-4020

ISBN 978-1-5108-1245-1

9 781510 812451

CURRAN ASSOCIATES INC.
proceedings
.com